Functional Statistics

Functional Statistics
Outliers Detection and Quality Control

Editor

Javier Martínez Torres

MDPI • Basel • Beijing • Wuhan • Barcelona • Belgrade • Manchester • Tokyo • Cluj • Tianjin

Editor
Javier Martínez Torres
Escuela Superior de Ingeniería y Tecnología,
Universidad Internacional de La Rioja
Spain

Editorial Office
MDPI
St. Alban-Anlage 66
4052 Basel, Switzerland

This is a reprint of articles from the Special Issue published online in the open access journal *Mathematics* (ISSN 2227-7390) (available at: https://www.mdpi.com/journal/mathematics/special_issues/Functional_Statistics_Outliers_Detection_Quality_Control).

For citation purposes, cite each article independently as indicated on the article page online and as indicated below:

LastName, A.A.; LastName, B.B.; LastName, C.C. Article Title. *Journal Name* **Year**, *Volume Number*, Page Range.

ISBN 978-3-03943-963-8 (Hbk)
ISBN 978-3-03943-964-5 (PDF)

© 2020 by the authors. Articles in this book are Open Access and distributed under the Creative Commons Attribution (CC BY) license, which allows users to download, copy and build upon published articles, as long as the author and publisher are properly credited, which ensures maximum dissemination and a wider impact of our publications.

The book as a whole is distributed by MDPI under the terms and conditions of the Creative Commons license CC BY-NC-ND.

Contents

About the Editor . . . . **vii**

Preface to "Functional Statistics" . . . . **ix**

Manuel Oviedo-de La Fuente, Celestino Ordóñez and Javier Roca-Pardiñas
Functional Location-Scale Model to Forecast Bivariate Pollution Episodes
Reprinted from: *Mathematics* **2020**, *8*, 941, doi:10.3390/math8060941 . . . . **1**

Ofélia Anjos, Miguel Martínez Comesaña, Ilda Caldeira, Soraia Inês Pedro, Pablo Eguía Oller and Sara Canas
Application of Functional Data Analysis and FTIR-ATR Spectroscopy to Discriminate Wine Spirits Ageing Technologies
Reprinted from: *Mathematics* **2020**, *8*, 896, doi:10.3390/math8060896 . . . . **13**

Ishaq Adeyanju Raji, Muhammad Hisyam Lee, Muhammad Riaz, Mu'azu Ramat Abujiya and Nasir Abbas
Outliers Detection Models in Shewhart Control Charts; An Application in Photolithography: A Semiconductor Manufacturing Industry
Reprinted from: *Mathematics* **2020**, *8*, 857, doi:10.3390/math8050857 . . . . **35**

Miguel Martínez Comesaña, Sandra Martínez Mariño, Pablo Eguía Oller, Enrique Granada Álvarez and Aitor Erkoreka González
A Functional Data Analysis for Assessing the Impact of a Retrofitting in the Energy Performance of a Building
Reprinted from: *Mathematics* **2020**, *8*, 547, doi:10.3390/math8040547 . . . . **53**

Elías Berriochoa, Alicia Cachafeiro, Alberto Castejón and José Manuel García-Amor
Classical Lagrange Interpolation Based on General Nodal Systems at Perturbed Roots of Unity
Reprinted from: *Mathematics* **2020**, *8*, 498, doi:10.3390/math8040498 . . . . **73**

Javier Martínez Torres, Jorge Pastor Pérez, Joaquín Sancho Val, Aonghus McNabola, Miguel Martínez Comesaña and John Gallagher
A Functional Data Analysis Approach for the Detection of Air Pollution Episodes and Outliers: A Case Study in Dublin, Ireland
Reprinted from: *Mathematics* **2020**, *8*, 225, doi:10.3390/math8020225 . . . . **91**

Miguel Flores, Salvador Naya, Rubén Fernández-Casal, Sonia Zaragoza, Paula Raña, Javier Tarrío-Saavedra
Constructing a Control Chart Using Functional Data
Reprinted from: *Mathematics* **2020**, *8*, 58, doi:10.3390/math8010058 . . . . **111**

About the Editor

Javier Martínez Torres has a PhD in Mathematics and Engineering from the University of Vigo. He is currently an Assistant Professor at the University of Vigo and has participated in more than 20 research projects as a principal investigator. He has published more than 50 papers in JCR indexed journals and participated in more than 25 international conferences.

Preface to "Functional Statistics"

Why use statistics and functional analysis? This is one of the most frequently asked questions with the rise of these types of techniques, especially when they are applied in areas other than mathematics, engineering, economics, etc. The reason is very straightforward: in this era of large amounts of data, most of the data available can be seen from a functional perspective, and thus, problems can be addressed through functional analysis and/or functional statistics. This book aims to build bridges between these two disciplines, which are sometimes far apart but are undoubtedly much closer than one imagines.

Javier Martínez Torres
Editor

Article

Functional Location-Scale Model to Forecast Bivariate Pollution Episodes

Manuel Oviedo-de La Fuente [1], Celestino Ordóñez [2,*] and Javier Roca-Pardiñas [3]

1. Department of Statistics, Mathematical Analysis and Optimization, Universidad de Santiago de Compostela, 15782 Santiago de Compostela, Spain; manuel.oviedo@usc.es
2. Department of Mining Exploitation and Propsecting, Universidad de Oviedo, Escuela Politécnica de Mieres, 33600 Mieres, Spain
3. Department of Statistics and Operation Research, Universidad de Vigo, 36310 Vigo, Spain; roca@uvigo.es
* Correspondence: ordonezcelestino@uniovi.es

Received: 7 May 2020; Accepted: 29 May 2020; Published: 8 June 2020

Abstract: Predicting anomalous emission of pollutants into the atmosphere well in advance is crucial for industries emitting such elements, since it allows them to take corrective measures aimed to avoid such emissions and their consequences. In this work, we propose a functional location-scale model to predict in advance pollution episodes where two pollutants are involved. Functional generalized additive models (FGAMs) are used to estimate the means and variances of the model, as well as the correlation between both pollutants. The method not only forecasts the concentrations of both pollutants, it also estimates an uncertainty region where the concentrations of both pollutants should be located, given a specific level of uncertainty. The performance of the model was evaluated using real data of SO_2 and NO_x emissions from a coal-fired power station, obtaining good results.

Keywords: pollution episodes; functional data; bivariate analysis; uncertainty region; generalized additive models

1. Introduction

Forecasting air quality and concentrations of pollutants in the atmosphere by means of statistical methods is an active area of research given the transcendence of the problem and the difficulty to find optimal solutions using deterministic mathematical models. Among the different methods that can be found in the literature to tackle this problem, models for time series analysis such as the integrated autoregressive moving average—ARIMA [1–3], multivariate regression [4–7], generalized linear or additive models (GAM) [8–11] and artificial neural networks (ANN) [12–19] are the most extended. Due to the increased access to continuous data over time, functional data analysis [20,21] was also proposed for air quality forecasting and outlier detection [22–24]. Parametric [25,26] and nonparametric [27–29] functional regression methods were tested. A functional framework allows considering the inherent correlation between observations, instead of considering them as independent realizations of an underlying stochastic process. Some functional approaches add related meteorological variables to the models [30–34], which can improve the result of the predictions and help to understand the process underlying the evolution of the pollutants.

Most of the documents in the literature propose solutions to predict the concentration of each pollutant individually, being much scarcer those focused on predicting more than one pollutant at a time. Vector autoregressive moving average (VARMA) [35,36] and vector autoregressive integrated moving average (VARIMA) [37] models were applied to reach this objective. In this work, we proposed a method for the simultaneous forecasting of pollution episodes when two pollutants, i.e., SO_2 and NO_x, are involved. Apart from transport, one of the main sources of these pollutants is public electricity and heating. Their negative effects on human health are well known, and goes for mild (i.e., eyes irritated, nose or headache) to severe (i.e., lung damage or reduced oxygenation of tissues). They also have negative effects on animals and plants, as well as in other substances, such as water and soils. In addition, NO_x is a precursor of the tropospheric ozone. High levels of ozone contributes to climate change, cause adverse impacts on health and can damage vegetation.

Pollution episodes (incidents) are abnormally large emissions of one or more pollutants in short periods of time. Although the improvement of the chemical processes and particle filter systems have significantly reduced the amount and intensity of the pollution episodes, they are still of particular interest for the industries, as they may be subject to sanctions, or for other reasons, such as public health deterioration or industry discredit. Therefore, pollution industries, such as coal-fired power plants, are very interested in determining in advance when these episodes of excessive contamination might occur. Specifically, this is the purpose of our work: forecasting pollution episodes of SO_2 and NO_x early enough to allow corrective measures to be taken. Our approach uses a location-scale model [11,38,39] that treats the predictors, the concentrations of both pollutants over time, as functions, while the response is a scalar, the concentration of the pollutants some time in advance. The novelty of our approach is the combination of a biviariate location-scale model with functional additive models. This method combines the simplicity of the location and scale models with the capacity of functional data analysis to deal with data in the form of functions.

The document is structured as follows: In Section 2 we show the mathematical model proposed to solve the problem under analysis and the algorithm used to estimate a solution from the data. Section 3 is devoted to test the validity of the model using real data. Finally, a discussion of the results and the main conclusions of our work are exposed in Section 4.

2. Methodology

2.1. Mathematical Model

Let $\{\mathbf{X}_i, \mathbf{Y}_i\}_{i=1}^n$ be a set of observations of a stochastic process, $\mathbf{X} = (X^1(t), \ldots, X^p(t))$, where $X^j(t) \in L_2[0, T]$, $j = 1, \ldots, p$, are predictor covariates and $\mathbf{Y} = (Y_1, Y_2)$, with $Y_j \in \mathbb{R}$, a response variable. In this context, the following bivariate location-scale model [40,41] is assumed

$$\begin{pmatrix} Y_1 \\ Y_2 \end{pmatrix} = \begin{pmatrix} \mu_1(\mathbf{X}) \\ \mu_2(\mathbf{X}) \end{pmatrix} + \mathbf{\Sigma}^{1/2}(\mathbf{X}) \begin{pmatrix} \varepsilon_1 \\ \varepsilon_2 \end{pmatrix} \quad (1)$$

where $\mathbf{\Sigma}^{1/2}(\mathbf{X})$ represents the Cholesky decomposition of the variance-covariance matrix $\mathbf{\Sigma}(\mathbf{X})$

$$\mathbf{\Sigma}(\mathbf{X}) = \begin{pmatrix} \sigma_1^2(\mathbf{X}) & \sigma_{12}(\mathbf{X}) \\ \sigma_{12}(\mathbf{X}) & \sigma_2^2(\mathbf{X}) \end{pmatrix} \quad (2)$$

so that $\text{Var}(\mathbf{Y}|\mathbf{X}) = \mathbf{\Sigma}(\mathbf{X}) = \mathbf{\Sigma}^{1/2}(\mathbf{X}) \left(\mathbf{\Sigma}^{1/2}(\mathbf{X})\right)^{\mathbf{T}}$. To guarantee the model identification in (1), the bivariate residuals $(\varepsilon_1, \varepsilon_2)$ are assumed to be independent of the covariates, with zero mean, unit variance, and zero correlation. Despite we do not assume any distribution for the error term, within the framework of functional data analysis this work might be addressed under the assumption of other structures for error distribution: generalized Gauss-Laplace distribution that relax the constrictive assumption of the normal distribution errors [42], generalized linear mixed models (GLMMs) [38] to estimate random effects and dependent (temporal or spatial) errors, and generalized additive models for location, scale and shape (GAMLSS) [43] to model the dynamically variable distribution, considering skewness and kurtosis.

We define the unconditionally probabilistic region for the errors $(\varepsilon_1, \varepsilon_2)$ as

$$\varepsilon_\tau(k) = \{(\varepsilon_1, \varepsilon_2) \in \mathbb{R}^2 | f(\varepsilon_1, \varepsilon_2) \geq k\}$$

f being the density function of the bivariate residuals $(\varepsilon_1, \varepsilon_2)$ and k the τ–quantile of $f(\varepsilon_1, \varepsilon_2)$. Then, for a given $\mathbf{X}$, we define a conditional τ^{th}- uncertainty region for (Y_1, Y_2) containing $\tau\%$ of the observations as:

$$R_\tau(\mathbf{X}) = \begin{pmatrix} \mu_1(\mathbf{X}) \\ \mu_2(\mathbf{X}) \end{pmatrix} + \Sigma^{1/2}(\mathbf{X}) \varepsilon_\tau \qquad (3)$$

2.2. Estimation Algorithm

To implement an algorithm that allows applying the mathematical model exposed in the previous section, we propose using a functional additive models to estimate the means, variances and covariances in (1). Given a sample of size n, $\{\mathbf{X}_i, (Y_{i1}, Y_{i2})\}_{i=1}^n$, where $\mathbf{X}_i = \left(X_i^1(t), \ldots, X_i^p(t)\right)$, the steps of the proposed estimation algorithm are the following:

Step 1: Perform a decomposition of each covariate $X^j(t)$ in basis functions of the form $X^j(t) \approx \sum_{k=1}^K \zeta_k^j \phi_k(t)$, where ϕ_k ($k = 1, \ldots, K$) are K basis functions (i.e., B-splines, wavelets), and ζ_{il} are either the coefficients of an expansion in fixed basis or the principal component scores of the Karhunen-Loève expansion [44,45]. As a result, we obtain the transformed covariates

$$\tilde{\mathbf{X}}_i = \left(\left(\zeta_{i1}^1, \ldots, \zeta_{iK}^1\right); \left(\zeta_{i1}^2, \ldots, \zeta_{iK}^2\right); \ldots; \left(\zeta_{i1}^p, \ldots, \zeta_{iK}^p\right) \right) \quad i = 1, \ldots, n$$

Step 2: For $r = 1, 2$, fit an additive model to the sample $\{\tilde{\mathbf{X}}_i, Y_{i1}, Y_{i2}\}_{i=1}^n$ and obtain an estimation of the means

$$\hat{\mu}_r(\mathbf{X}_i) = \alpha_r + \sum_{j=1}^p \sum_{k=1}^K \hat{f}_{rk}^j(\zeta_{ik}^j) \qquad (4)$$

and then estimate $\sigma_r^2(\mathbf{X})$ from the sample $\{\tilde{\mathbf{X}}_i, (Y_{ir} - \hat{\mu}_r(\mathbf{X}_i))^2\}_{i=1}^n$ as

$$\hat{\sigma}_r^2(\mathbf{X}_i) = \exp\left(\hat{\beta}_r + \sum_{j=1}^p \sum_{k=1}^K \hat{g}_{rk}^j(\zeta_{ik}^j) \right) \qquad (5)$$

Then, compute the correlation $\rho(\mathbf{X})$, which is related to the covariance by $\sigma_{12}(\mathbf{X}) = \sigma_1(\mathbf{X})\sigma_2(\mathbf{X})\rho(\mathbf{X})$, using the sample $\{\mathbf{X}_i, \hat{\delta}_i\}_{i=1}^n$, as follows:

$$\hat{\rho}(\mathbf{X}_i) = \tanh\left(\hat{\gamma} + \sum_{j=1}^p \sum_{k=1}^K \hat{m}_k^j(\zeta_{ik}^j) \right)$$

being

$$\hat{\delta}_i = \frac{(Y_i^1 - \hat{\mu}_1(\mathbf{X}_i))(Y_i^2 - \hat{\mu}_2(\mathbf{X}_i))}{\hat{\sigma}_1(\mathbf{X}_i)\hat{\sigma}_2(\mathbf{X}_i)}$$

where f_{rk}^j, g_{rk}^j and m_k^j are smooth and unknown functions, α_r, β_r and γ are coefficients, p the number of predictors (covariates), and K the number of basis. Please note that the link functions $H_\sigma(\cdot) = \exp(\cdot)$ and $H_\rho(\cdot) = \tanh(\cdot)$ used in the variance and correlation structures, respectively, ensure that the restrictions on the parameter spaces ($\sigma_r^2(\mathbf{X}) \geq 0$ and $0 \leq \rho(\mathbf{X}) \leq 1$) are maintained. Moreover, in order to guarantee the identification of the model, we assume that all the means of functions f_j, g_j and m_j are zero.

Step 3: Compute the standardized residuals

$$\begin{pmatrix} \hat{\varepsilon}_{i1} \\ \hat{\varepsilon}_{i2} \end{pmatrix} = \hat{\Sigma}^{-1/2}(\mathbf{X}_i) \begin{pmatrix} Y_{i1} - \hat{\mu}_1(\mathbf{X}_i) \\ Y_{i2} - \hat{\mu}_2(\mathbf{X}_i) \end{pmatrix} \quad i = 1, \ldots, n \qquad (6)$$

where

$$\hat{\Sigma}(\mathbf{X}_i) = \begin{pmatrix} \hat{\sigma}_1^2(\mathbf{X}_i) & \hat{\sigma}_{12}(\mathbf{X}_i) \\ \hat{\sigma}_{12}(\mathbf{X}_i) & \hat{\sigma}_2^2(\mathbf{X}_i) \end{pmatrix} \quad (7)$$

with $\hat{\sigma}_{12}(\mathbf{X}_i) = \hat{\sigma}_1(\mathbf{X}_i)\hat{\sigma}_2(\mathbf{X}_i)\hat{\rho}(\mathbf{X}_i)$.

Step 4: Obtain the kernel estimation of the bivariate density $\hat{f}(\varepsilon_1, \varepsilon_2)$ given by

$$\hat{f}((\varepsilon_1, \varepsilon_2), \mathbf{H}) = \frac{1}{n} \sum_{i=1}^{n} K_\mathbf{H} \begin{pmatrix} \varepsilon_1 - \hat{\varepsilon}_{i1} \\ \varepsilon_2 - \hat{\varepsilon}_{i2} \end{pmatrix} \quad (8)$$

where $K(\cdot)$ is the kernel which is a symmetric probability density function and $\mathbf{H}$ is a 2×2 positive definite matrix. Then, obtain the τth unconditional bivariate uncertainty region on the residual scale as

$$\hat{\varepsilon}_\tau = \{(\varepsilon_1, \varepsilon_2)) \in \mathbb{R}^2 | \hat{f}(\varepsilon_1, \varepsilon_2)) \geq \hat{k}\} \quad (9)$$

$\hat{k}$ being the empirical τ quantile of the values $\hat{f}(\varepsilon_{11}, \varepsilon_{12}), \ldots, \hat{f}(\varepsilon_{n1}, \varepsilon_{n2})$. Finally, for a given $\mathbf{X}$, the conditional bivariate uncertainty region $R_\tau(\mathbf{X})$ is estimated according to (3)

$$\hat{R}_\tau(\mathbf{X}) = \begin{pmatrix} \hat{\mu}_1(\mathbf{X}) \\ \hat{\mu}_2(\mathbf{X}) \end{pmatrix} + \hat{\Sigma}^{1/2}(\mathbf{X})\hat{\varepsilon}_\tau \quad (10)$$

3. Case Study: Joint Forecasting of (SO_2, NO_x) Pollution Episodes

The mathematical model exposed in the previous section was applied to the forecasting of pollution episodes registered at a coal-fired power station located in the northwest of Spain. SO_2 and NO_x are two of the main air pollutants generated by combustion processes, and both have harmful effects on human health. Moreover, it was proven that both pollutants are correlated [46], which is consistent with the model in (1). Fortunately, pollution episodes are not very frequent and the trend is that they will become scarcer as technology advances.

Let t_0 be the present time measured each five minutes, and $SO(t_0)$ and $NO(t_0)$ the concentrations obtained respectively by the series of bi-hourly SO_2 and NO_x means at instant t_0. Being t_h the prediction horizon time, the interest is to predict

$$(Y_1, Y_2) = (SO(t_0 + t_h), NO(t_0 + t_h))$$

and provide an uncertainty region for these estimations given a specific value of τ, using the predictive covariates

$$\mathbf{X} = \left(X^1(t), X^2(t), X^3(t), X^4(t)\right) = (SO(t), NO(t), SO'(t), NO'(t)) \quad \text{with} \quad t \in [t_0 - t_{lag}, t_0]$$

where $(NO'(t), SO'(t))$ represents the first derivatives of the functions that approximate the concentrations of both pollutants. These derivatives are obtained from the functional representation of the discrete data, according to Step 1 of the estimation algorithm. Please note that t_{lag} represents the lagged time used in the predictors. In particular,, we are interested in predicting an hour in advance, according to the requirements of current Spanish legislation and, therefore, we will consider $t_h = 12$ (60 min) from now on.

Most of the time, these concentrations times series are low, close to zero, and in order to obtain a reasonably large number of pollution incidents, we took as our sample a historical matrix $\{(\mathbf{X}_i, \mathbf{Y}_i)\}_{i=1}^{N}$ with pollution data of approximately 12 years, which includes a considerable number of pollution episodes (see [9] for a detailed description of the historical matrix construction). In summary, in the historical matrix not all the data are used, but only part of them, following a quantile-weighted criterion. This means that the larger the concentration, the greater the number of observations of that concentration in the sample. Figure 1 shows a sample of the historical matrix, on top are the curves of both pollutants $(NO(t), SO(t))$ measured in $t_{lag} = 20$ discretization points and evaluated in

5 B-spline basis functions of order $p = 4$. On the bottom, the first derivative of the B-spline curves $(\text{NO}'(t), \text{SO}'(t))$ with order $p - 1$ are represented.

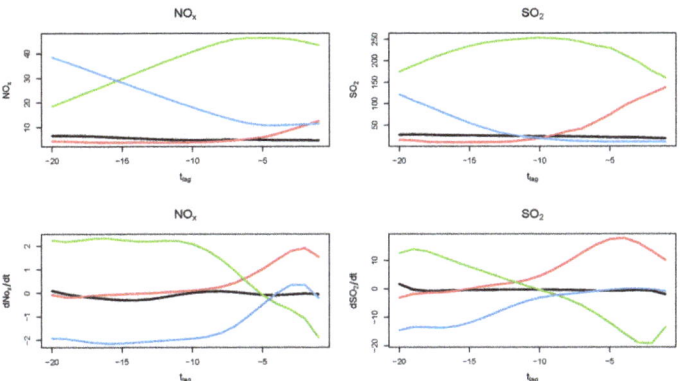

Figure 1. Five curves of the historical matrix for pollutants $(\text{NO}(t), \text{SO}(t))$ and their corresponding first derivatives $(\text{NO}'(t), \text{SO}'(t))$, observed in a period of time $t_{lag} = 20$.

In this paper, we tested four models using as predictors different combinations of the covariates $(X^1(t), X^2(t), X^3(t), X^4(t))$ that include the concentrations of both pollutants and their first derivatives. In particular, we will consider models given by:

$$\begin{pmatrix} SO(t_0 + 12) \\ NO(t_0 + 12) \end{pmatrix} = \begin{pmatrix} \mu_1(\mathbf{X}^1) \\ \mu_2(\mathbf{X}^2) \end{pmatrix} + \begin{pmatrix} \sigma_1^2(\mathbf{X}^1) & \sigma_{12}(\mathbf{X}^1, \mathbf{X}^2) \\ \sigma_{12}(\mathbf{X}^1, \mathbf{X}^2) & \sigma_2^2(\mathbf{X}^2) \end{pmatrix} \begin{pmatrix} \varepsilon_1 \\ \varepsilon_2 \end{pmatrix} \quad (11)$$

The four considered models, M_1, M_2, M_3 and M_4, are configured in Table 1 where the cross X indicates the covariates included in each model.

Table 1. Selected models from equation in (11). Cross X indicates the covariates included in each of the four considered models. The derivatives of the functions are indicated with a single quote.

	X^1				X^2			
Model	SO(t)	NO(t)	SO'(t)	NO'(t)	SO(t)	NO(t)	SO'(t)	NO'(t)
M_1	X					X		
M_2	X	X			X	X		
M_3	X		X			X		X
M_4	X	X	X		X	X	X	X

To validate and compare the four proposed models, we randomly select from the full historical matrix a training set $\mathbf{M}^I = \{(\mathbf{X}_i^I, \mathbf{Y}_i^I)\}_{i=1}^{n_{train}}$ and a test set $\mathbf{M}^{II} = \{(\mathbf{X}_i^{II}, \mathbf{Y}_i^{II})\}_{i=n_{train}+1}^{N}$.

The estimates $\hat{\mu}_1$, $\hat{\mu}_2$, $\hat{\boldsymbol{\Sigma}}$ were obtained from the samples in the first matrix $\mathbf{M}_i^I$. The bivariate uncertainty regions for the values of the covariates on the second matrix $\mathbf{M}^{II}$ were obtained using (3). The estimated coverage $\hat{\tau}$ is given by

$$\hat{\tau} = \frac{1}{n_{test}} \sum_{i=n_{train}+1}^{N} I\{\mathbf{Y}_i^{II} \in \hat{R}_\tau(\mathbf{X}_i^{II})\}; n_{test} = N - n_{train} \quad (12)$$

The performance of the proposed predictors was evaluated in two pollution incidents. A bivariate representation of these episodes is shown in Figure 2. The orientation of the points shows a clear correlation between both pollutants although the range of concentrations is quite different.

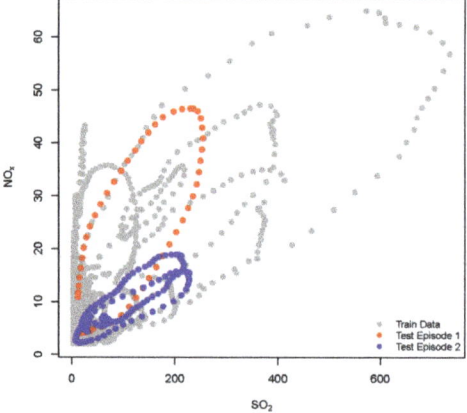

Figure 2. Observed and forecasted concentrations of SO_2 and NO_x for two pollution episodes.

The nominal and the estimated coverages for different time lags and training sample sizes are shown in Table 2. The coverages correspond to the bivariate solution, and were obtained for n_{test} consecutive observations that might or might not correspond to pollution incidents.

Table 2. Nominal τ and estimated $\hat{\tau}$ coverages for each of the four models under study. Two time lags, $t_{lag} = 10$, $t_{lag} = 20$, two sizes of the training sample $n_{train} = 10{,}000$ and $n_{train} = 4900$, and two numbers of principal components, K = 3 and K = 5, were considered. Results correspond to the test sample.

				$\hat{\tau}$			
τ	t_{lag}	$n^{\bullet}_{train}$	K	M_1	M_2	M_3	M_4
0.50	10	20	3	0.43	0.47	0.45	0.51
			5	0.42	0.48	0.47	0.49
	10	49	3	0.51	0.52	0.52	0.52
			5	0.51	**0.50**	**0.50**	**0.50**
	20	20	3	0.45	0.49	0.44	0.49
			5	0.48	0.46	0.43	0.46
	20	49	3	0.50	0.54	0.51	0.50
			5	0.49	0.49	0.49	0.48
0.75	10	20	3	0.70	0.73	0.72	**0.75**
			5	0.69	0.74	0.73	0.74
	10	49	3	0.76	0.78	0.78	0.78
			5	0.77	0.76	0.76	0.76
	20	20	3	0.70	0.72	0.70	0.72
			5	0.70	0.71	0.69	0.69
	20	49	3	0.77	0.78	**0.75**	0.73
			5	**0.75**	0.74	0.72	0.72
0.90	10	20	3	0.88	0.87	0.87	0.89
			5	0.86	0.88	0.87	0.88
	10	49	3	0.91	**0.90**	**0.90**	**0.90**
			5	**0.90**	**0.90**	**0.90**	**0.90**
	20	20	3	0.87	0.87	0.85	0.86
			5	0.87	0.86	0.86	0.84
	20	49	3	0.90	0.89	0.90	0.86
			5	0.87	0.87	0.88	0.85
0.95	10	20	3	0.93	0.93	0.93	0.93
			5	0.93	0.93	0.93	0.93
	10	49	3	0.96	0.93	0.93	0.93
			5	0.95	0.94	0.94	0.94
	20	20	3	0.93	0.93	0.92	0.92
			5	0.92	0.92	0.93	0.90
	20	49	3	**0.95**	0.94	**0.95**	0.92
			5	0.92	0.92	0.93	0.90

RMSE values for each model are shown in Table 3, considering an expansion of the functions in three or five principal components, and lags $t_{lag} = 10$ and $t_{lag} = 20$. Please note that this table makes reference to the marginal distributions, and that the range of concentrations for each pollutant is very different, therefore the RMSE values are also different.

Table 3. RMSE values for two pollution episodes and the four models tested, considering curves with two different time lags, size of the training samples and number of principal components.

Response	t_{lag}	$n^{\bullet}_{train}$	K	Episode 1				Episode 2			
				M_1	M_2	M_3	M_4	M_1	M_2	M_3	M_4
NO$_x$	10	20	3	20.7	25.2	20.0	23.0	1.9	1.7	1.4	1.4
			5	19.6	18.6	20.6	**16.4**	0.9	0.8	0.8	**0.6**
		49	3	20.9	24.7	18.2	23.8	1.8	1.9	1.4	1.4
			5	19.2	17.5	19.3	16.7	0.9	0.9	0.8	0.7
	20	100	3	34.0	24.7	19.9	19.6	3.3	5.5	1.7	3.1
			5	40.9	23.3	46.5	29.2	1.5	2.5	1.0	1.8
		49	3	30.2	18.2	18.7	20.4	3.4	5.3	1.7	2.8
			5	36.2	27.6	39.8	35.6	1.5	2.5	1.0	1.8
SO$_2$	10	100	3	505.0	841.0	**407.5**	837.3	544.8	531.9	419.8	419.1
			5	914.9	868.6	669.4	516.3	215.6	230.3	184.5	199.1
		49	3	686.0	685.3	515.8	518.5	481.4	484.4	338.0	361.1
			5	991.1	925.5	846.3	682.9	199.9	214.7	**170.6**	179.9
	20	100	3	1463.4	2172.7	825.4	1199.9	1154.5	1133.1	709.5	659.0
			5	1470.6	2531.2	1002.9	1428.5	525.5	482.1	352.0	341.3
		49	3	1.458.7	2485.4	768.4	698.4	1162.6	1125.8	644.8	628.7
			5	1787.6	2811.0	1111.2	951.5	548.1	492.0	352.7	359.3

Note: $n_{train} = 100 \cdot n^{\bullet}_{train}$.

For the two episodes analyzed, Figure 3 shows the observed and the predicted values as well as the quantile for $\hat{\tau} = 0.95$, calculated for the test sample. The results correspond to curves observed in ten points ($t_{lag} = 10$) and represented in a basis expansion in three functional principal components. These univariate confidence intervals were respectively constructed from (11) as $\mu_1(\mathbf{X}^1) + \sigma_1(\mathbf{X}^1)\varepsilon_1^{0.95}$ and $\mu_2(\mathbf{X}^2) + \sigma_2(\mathbf{X}^2)\varepsilon_2^{0.95}$, $\varepsilon_1^{0.95}$ and $\varepsilon_1^{0.95}$ being the 0.95 quantile of the distributions of errors ε_1 and ε_2, respectively.

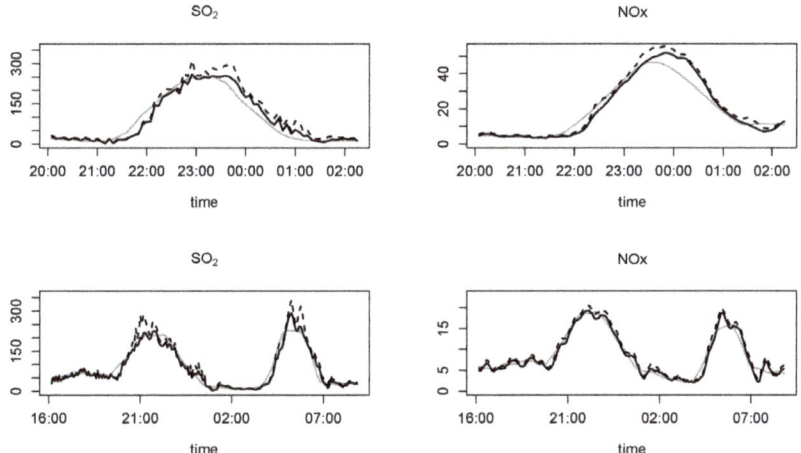

Figure 3. Observations (solid black line), mean (solid gray line) and 0.95th quantile estimations (discontinuous line) for both pollutants, SO$_2$ and NO$_x$ and for two pollution episodes.

Table 4 shows the maximum consumed memory and the runtime (in seconds) for the four models tested and two different dimensions of the submatrices are executed in a Intel Core i7-2600K with 16 GB of RAM.

Table 4. Maximum memory consumption (MB) and computation time (seconds) for the four models tested, M_i, following different strategies concerning the time lag, t_{lag}, the size of the training sample, n_{train}, and the number of basis functions, K.

			Memory (MB)				Runtime (seconds)			
t_{lag}	n_{train}	K	M_1	M_2	M_3	M_4	M_1	M_2	M_3	M_4
10	20	3	652.70	1083.03	1319.57	2266.34	17.98	29.66	35.99	61.62
		5	1090.78	1855.39	2299.14	4075.88	35.55	55.29	63.15	124.7
	49	3	329.94	548.58	669.42	1153.38	10.53	17.42	21.38	37.01
		5	552.74	942.68	1171.67	2089.53	19.79	31.28	46.62	88.20
20	20	3	653.34	1084.15	1320.66	2268.16	18.11	29.63	35.97	61.21
		5	1091.90	1857.26	2300.99	4078.95	32.14	49.51	72.45	124.81
	49	3	330.10	548.84	669.69	1153.82	10.51	17.56	21.49	37.82
		5	553.01	943.13	1172.20	2090.36	17.80	36.44	39.03	76.11

Note: $n_{train} = 100 \cdot n^{\bullet}_{train}$.

4. Discussion

We begin the discussion of the results analyzing Table 2 that show the estimated coverage for the bivariate prediction depending on the time lag, the size of the training sample, the number of principal components and the model used. It can be appreciated that the estimated coverages are generally lower than the theoretical coverages, although very close. Therefore, the mathematical models proposed show a good performance although there is a trend to underestimate the observed values. This effect can also be appreciated in Figure 3, where the mean tends to be under the observed values. Then, in order to be on the safe wide, it would be preferable to use the quantile $\tau = 0.95$, which provides greater guarantee of predicting the highest values of the pollution episodes. Regarding the rest of the parameters, it is not possible to establish a combination of them that provides the best results. However, in general they were obtained for the lowest training size, $n_{train} = 49,000$, and for models that includes one or two derivatives (models M_3 and M_4).

The prediction errors are shown in Table 1, where the best results (minimum RMSE) are marked in bold. As can be seen, they correspond to model M_4 for NO_x and model M_3 for SO_2. In both cases, these models incorporate the derivatives of the original functions. Accordingly, we conclude that the derivatives contribute positively to improve the results, which reinforces the role of the functional approach. However, there is an asymmetry between both pollutants: using the concentrations of SO_2 and their derivatives improves the results for NO_x, but using the concentrations of NO_x and their derivatives is not an advantage in the estimation of SO_2. When SO_2 and NO_x concentrations of both episodes are plotted against time (Figure 4), a slight advance can be seen on the first pollutant compared to the second, which would explain this asymmetry.

With respect to the time lag, the minimum RMSE values were obtained for the shorter period of time $t_{lag} = 10$, so it seems that using 20 observations to predict one our in advance introduced noise into the model instead of adding useful information. This result is in agreement with those obtained for the same data in previous studies of some of the authors, which indicated that only a few observations close to the time of prediction contribute to that prediction. Talking about the size of the training sample, simplifying the original data by removing small values of the concentrations improves the results in most of the cases, so this would be the advisable option.

When the effect of the number of principal components used as basis functions is analyzed, using $K = 5$ is always favorable for episode 2, for both SO_2 and NO_x, but not for episode 1, for which the trend is opposite.

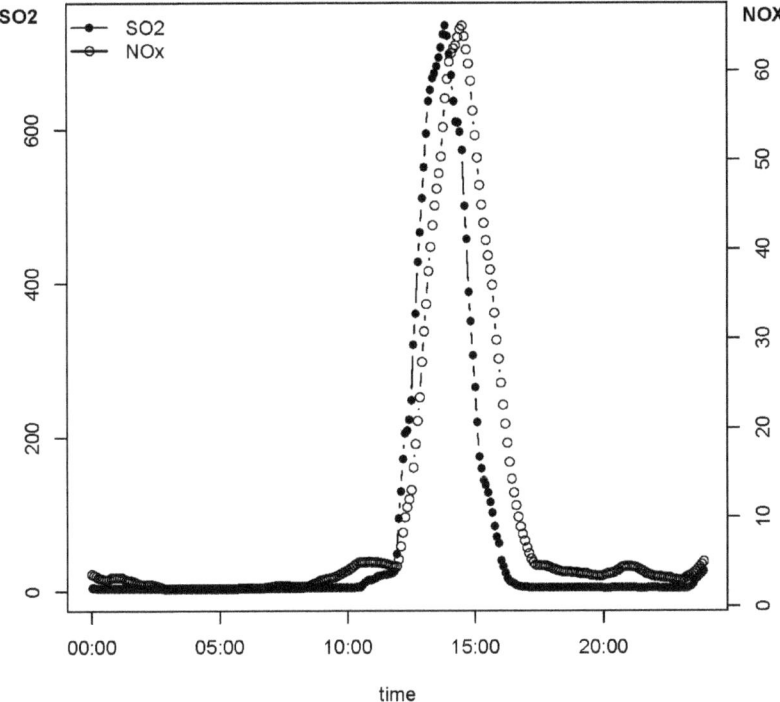

Figure 4. Example of a pollution incident showing SO_2 and NO_x concentrations versus time. Notice that there is an advance in the first pollutant compared to the second.

Although they are not shown in the article, so as not to overstretch it, a comparison of the estimated coverages using 3 or 5 principal components, or 5 B-splines basis functions, tell us that there are not substantial differences among them, so it seems that one or other base functions can be used interchangeably.

Finally, regarding memory consumption and runtime for model training, it is evident, from Table 4 that more complex models consume more resources and requires more computing time. For fixed values of the time lag (t_{lag}) and the size of the training sample (n_{train}), model M_4 is between 3 and 7 times more expensive than model M_1 in terms of memory consumption and runtime. Using time lags $t_{lag} = 10$ or $t_{lag} = 50$ has no effect in terms of computation requirements; and employing 5 principal components instead of 3 principal components implies an approximately double memory consumption and runtime.

To conclude, it is possible to establish that the functional location-scale model proposed were quite a good approach (in terms of coverage and prediction error) to forecast bivariate pollution episodes one hour in advance, as it is required by the Spanish legislation. The best results were obtained when the derivatives of functions adjusted to the observed data are included in the model, when the raw data are filtered and when the shorter period of time is used for the prediction. The size of the training data and the type and number of basis functions are, instead, parameters on which definitive conclusions could not be drawn.

Author Contributions: Conceptualization, J.R.-P. and C.O.; methodology, J.R.-P. and M.O.-d.L.F.; software, J.R.-P. and M.O.-d.L.F.; validation, J.R.-P., C.O. and M.O.-d.L.F.; writing—original draft preparation, J.R.-P. and C.O.; writing—review and editing, J.R-P., C.O. and M.O.-d.L.F.; supervision, C.O. All authors have read and agreed to the published version of the manuscript.

Acknowledgments: The authors acknowledge financial support from: (1) UO-Proyecto Uni-Ovi (PAPI-18-GR-2014-0014), (2) Project MTM2016-76969-P from Ministerio de Economía y Competitividad—Agencia Estatal de Investigación and European Regional Development Fund (ERDF) and IAP network StUDyS from Belgian Science Policy, (3) Nuevos avances metodológicos y computacionales en estadística no-paramétrica y semiparamétrica—Ministerio de Ciencia e Investigación (MTM2017-89422-P).

Conflicts of Interest: The authors declare no conflict of interest.

References

1. Siew, L.Y.; Ching, L.Y.; Wee, P.M.J. ARIMA and integrated ARFIMA models for forecasting air pollution index in Shah Alam, Selangor. *Malay. J. Anal. Sci.* **2008**, *12*, 257–263.
2. Ibrahim, M.Z.; Roziah, Z.; Marzuki, I.; Muhd S.L. Forecasting and Time Series Analysis of Air Pollutants in Several Area of Malaysia. *Am. J. Enverion. Sci.* **2009**, *5*, 625–632. [CrossRef]
3. Abhilash, M.S.K.; Thakur, A.; Gupta, D.; Sreevidya, B. Time Series Analysis of Air Pollution in Bengaluru Using ARIMA Model. In *Ambient Communications and Computer Systems*; Advances in Intelligent Systems and Computing; Perez, G., Tiwari, S., Trivedi, M., Mishra, K., Eds.; Springer: Singapore, 2018.
4. Liu, P.W.G. Simulation of the daily average PM10 concentrations at Ta-Liao with Box-Jenkins time series models and multivariate analysis. *Atmos. Environ.* **2009**, *43*, 2104–2113. [CrossRef]
5. Nazif, A.; Mohammed, N.I.; Malakahmad, A.; Abualqumboz, M.S. Regression and multivariate models for predicting particulate matter concentration level. *Environ. Sci. Pollut. Res. Int.* **2018**, *25*, 283–289. [CrossRef]
6. Zhao, R.; Gu, X.; Xue, B.; Zhang, J.; Ren, W. Short period PM2.5 prediction based on multivariate linear regression model. *PLoS ONE* **2018**, *13*, e0201011. [CrossRef]
7. Ng, K.Y.; Awang, N. Multiple linear regression and regression with time series error models in forecasting PM10 concentrations in Peninsular Malaysia. *Environ. Monit. Assess.* **2018**, *190*, 63. [CrossRef]
8. Roca-Pardiñas, J.; Gonzàlez Manteiga, W.; Febrero-Bande, M.; Prada-Sànchez, J.M.; Cadarso-Suàrez, C. Predicting binary time series of SO_2 using generalized additive models with unknown link function. *Environmetrics* **2004**, *15*, 729–742. [CrossRef]
9. Martínez-Silva, I.; Roca-Pardiñas, J.; Ordóñez, C. Forecasting SO_2 pollution incidents by means of quantile curves based on additive models. *Environmetrics* **2016**, *27*, 147–157. [CrossRef]
10. Garcia, J.M.; Teodoro F.; Cerdeira, R.; Coelho, L.M.R.; Prashant, K.; Carvalho, M.G. Developing a methodology to predict PM10 concentrations in urban areas using generalized linear models. *Environ. Technol.* **2016**, *37*, 2316–2325. [CrossRef]
11. Roca-Pardiñas, J.; Ordóñez, C. Predicting pollution incidents through semiparametric quantile regression models. *Stoch. Environ. Res. Risk Assess.* **2019**, *33*, 673–685. [CrossRef]
12. Azid, I.A.; Ripin, Z.M.; Aris, M.S.; Ahmad, A.L.; Seetharamu, K.N.; Yusoff, R.M. Predicting combined-cycle natural gas power plant emissions by using artificial neural networks. In Proceedings of the 2000 TENCON Proceedings, Intelligent Systems and Technologies for the New Millennium (Cat. No.00CH37119), Kuala Lumpur, Malaysia, 24–27 September 2000; Volume 3, pp. 512–517.
13. Perez, P.; Trier, A.; Reyes, J. Prediction of PM2.5 Concentrations Several Hours in Advance Using Neural Networks in Santiago, Chile. *Atmos. Environ.* **2000**, *34*, 1189–1196. [CrossRef]
14. Ferretti, G.; Piroddi, L. Estimation of NO_x Emissions in Thermal Power Plants Using Neural Networks. *J. Eng. Gas Turbines Power* **2001**, *132*, 465–471. [CrossRef]
15. Siwek, K.; Osowski, S. Improving the accuracy of prediction of PM10 pollution by the wavelet transformation and an ensemble of neural predictors. *Eng. Appl. Artif. Intell.* **2012**, *25*, 1246–1258. [CrossRef]
16. Muñoz, E.; Martín, M.L.; Turias, I.J.; Jimenez-Come, M.J.; Trujillo, F.J. Prediction of PM10 and SO_2 exceedances to control air pollution in the Bay of Algeciras, Spain. *Stoch. Environ. Res. Risk Assess.* **2014**, *28*, 1409–1420. [CrossRef]
17. He, H.D.; Lu, W.Z.; Xue, Y. Prediction of particulate matters at urban intersection by using multilayer perceptron model based on principal components. *Stoch Environ. Res. Risk. Assess.* **2015**, *29*, 2107–2114. [CrossRef]
18. Antanasijević, D.; Pocajt, V.; Perić-Grujić, A.; Ristić, M. Multiple-input–multiple-output general regression neural networks model for the simultaneous estimation of traffic-related air pollutants. *Atmos. Pollut. Res.* **2018**, *9*, 388–397. [CrossRef]

19. Gilson, M.; Dahmen, D.; Moreno-Bote, R.; Insabato, A.; Helias M. The covariance perceptron: A new paradigm for classification and processing of time series in recurrent neuronal networks. *BioRxiv* **2019**. [CrossRef]
20. Ramsay, J.O.; Silverman, B.W. *Applied Functional Data Analysis: Methods and Case Studies*; Springer: New York, USA, 2002.
21. Ferraty, F.; Vieu, P. *Nonparametric Functional Data Analysis*; Springer: New York, USA, 2006.
22. Febrero-Bande, M.; Galeano, P.; González-Manteiga, W. Outlier detection in functional data by depth measures with application to identify abnormal NOx levels. *Environmetrics* **2008**, *19*, 331–345. [CrossRef]
23. Martinez, J.; Saavedra, Á.; García-Nieto, P.J.; Piñeiro, J.I.; Iglesias, C.; Taboada, J.; Sanchoa, J.; Pastor, J. Air quality parameters outliers detection using functional data analysis in the Langreo urban area (Northern Spain). *Appl. Math. Comput.* **2014**, *241*, 1–10. [CrossRef]
24. Shaadan, N.; Jemain, A.A.; Latif, M.T. Anomaly detection and assessment of PM10, functional data at several locations in the Klang Valley, Malaysia. *Atmos. Pollut. Res.* **2015**, *6*, 365–375. [CrossRef]
25. Ignaccolo, R.; Mateu, J.; Giraldo, R. Kriging with external drift for functional data for air quality monitoring. *Stoch. Environ. Res. Risk Assess.* **2014**, *28*, 1171–1186. [CrossRef]
26. Wang, D.; Zhong, Z.; Kaixu, B.; Lingyun, H. Spatial and Temporal Variabilities of PM2.5 Concentrations in China Using Functional Data Analysis. *Sustainability* **2019**, *11*, 1620. [CrossRef]
27. Aneiros-Pérez, G.; Cardot, H.; Estévez-Pérez, G.; Vieu, P.H. Maximum ozone concentration forecasting by functional non-parametric approaches. *Environmetrics* **2004**, *15*, 675–685. [CrossRef]
28. Fernández de Castro, B.M.; González-Manteiga, W.; Guillas, S. Functional samples and bootstrap for predicting sulfur dioxide levels. *Technometrics* **2005**, *47*, 212–222. [CrossRef]
29. Quintela-del-Río, A.; Francisco-Fernández, M. Nonparametric functional data estimation applied to ozone data: Prediction and extreme value analysis. *Chemosphere* **2001**, *82*, 800–808. [CrossRef]
30. Besse, P.C.; Cardot, H.; Stephenson, D.B. Autoregressive forecasting of some functional climatic variations. *Scand. J. Stat.* **2000**, *27*, 673–687. [CrossRef]
31. Damon, J.; Guillas, S. The inclusion of exogenous variables in functional autoregressive ozone forecasting. *Environmetrics* **2002**, *13*, 759–774. [CrossRef]
32. Ruiz-Medina, M.D.; Espejo R.M. Spatial autoregressive functional plug-in prediction of ocean surface temperature. *Stoch. Environ. Res. Risk. Assess.* **2012**, *26*, 335–344. [CrossRef]
33. Ruiz-Medina, M.D.; Espejo, R.M.; Ugarte, M.D.; Militino A.F. Functional time series analysis of spatio-temporal epidemiological data. *Stoch. Environ. Res. Risk Assess.* **2014**, *28*, 943–954. [CrossRef]
34. Alvarez-Liebana, J.; Ruiz Medina, M.D. Prediction of air pollutants PM10 by ARBX(1) processes. *Stoch. Environ. Res. Risk Assess.* **2019**, *33*, 1721–1736. [CrossRef]
35. Hsu K.J. Time series analysis of the interdependence among air pollutants. *Atm. Environ. Part B Urban Atmos.* **1992**, *26*, 491–503. [CrossRef]
36. Kadiyala, A.; Kumar, A. Vector time series models for prediction of air quality inside a public transportation bus using available software. *Environ. Prog. Sustain.* **2014**, *33*, 337–341. [CrossRef]
37. García-Nieto, P.J.; Sánchez-Lasheras, F.; García-Gonzalo, E.; de Cos Juez F.J. Estimation of PM10 concentration from air quality data in the vicinity of a major steelworks site in the metropolitan area of Avilés (Northern Spain) using machine learning techniques. *Stoch Environ. Res. Risk Assess.* **2018**, *32*, 3287–3298. [CrossRef]
38. Hedeker, D.; Mermelstein, R.J.; Demirtas, H. An Application of a Mixed-Effects Location Scale Model for Analysis of Ecological Momentary Assessment (EMA) Data. *Biometrics* **2008**, *64*, 627–634. [CrossRef]
39. Taylor, J.; Verbyla, A. Joint modelling of location and scale parameters of the t distribution. *Stat. Model.* **2004**, *4*, 91–112. [CrossRef]
40. Pugach, O.; Hedeker, D.; Mermelstein, R. A Bivariate Mixed-Effects Location-Scale Model with application to Ecological Momentary Assessment (EMA) data. *Health Serv. Outcomes Res. Methodol.* **2014**, *14*, 194–212. [CrossRef]
41. He, W.; Lawless, J.F. Bivariate location-scale models for regression analysis, with applications to lifetime data. *J. R. Statist. Soc. B* **2005**, *67 Pt 1*, 63–78. [CrossRef]
42. Jäntschi, L.; Bálint, D.; Bolboaca, S.D. Multiple Linear Regressions by Maximizing the Likelihood under Assumption of Generalized Gauss-Laplace Distribution of the Error. *Comput. Math. Methods Med.* **2016**, *2016*, 8578156. [CrossRef]

43. Rigby, R.A.; Stasinopoulos, D.M. Generalized additive models for location, scale and shape. *J. R. Stat. Soc. Ser. C* **2005**, *54*, 507–554. [CrossRef]
44. Karhunen, K. Zur Spektraltheorie Stochastischer Prozesse. *Annales Academiae Scientiarum Fennicae Series A1 Mathematica-Physica* **1946**, *54*, 1–7. Available online: https://katalog.ub.uni-heidelberg.de/cgi-bin/titel.cgi?katkey=67295489 (accessed on 24 February 2020).
45. Febrero-Bande, M.; Oviedo de la Fuente, M. Statistical Computing in Functional Data Analysis: The R Package fda.usc. *J. Stat. Softw.* **2012**, *51*, 12. [CrossRef]
46. Dogruparmak, S.C.; Özbay, B. Investigating Correlations and Variations of Air Pollutant Concentrations under Conditions of Rapid Industrialization–Kocaeli (1987–2009). *Clean-Soil Air Water* **2011**, *39*, 597–604. [CrossRef]

© 2020 by the authors. Licensee MDPI, Basel, Switzerland. This article is an open access article distributed under the terms and conditions of the Creative Commons Attribution (CC BY) license (http://creativecommons.org/licenses/by/4.0/).

Article

Application of Functional Data Analysis and FTIR-ATR Spectroscopy to Discriminate Wine Spirits Ageing Technologies

Ofélia Anjos [1,2,3,*], Miguel Martínez Comesaña [4], Ilda Caldeira [5,6], Soraia Inês Pedro [1], Pablo Eguía Oller [4] and Sara Canas [5,6]

1. Instituto Politécnico de Castelo Branco, Escola Superior Agrária, 6001-909 Castelo Branco, Portugal; soraia_p1@hotmail.com
2. CEF, Instituto Superior de Agronomia, Universidade de Lisboa, 1349-017 Lisboa, Portugal
3. CBPBI, Centro de Biotecnologia de Plantas da Beira Interior, 6001-909 Castelo Branco, Portugal
4. Department of Mechanical Engineering, Heat Engines and Fluid Mechanics, Industrial Engineering School, University of Vigo, Maxwell s/n, 36310 Vigo, Spain; migmartinez@uvigo.es (M.M.C.); peguia@uvigo.es (P.E.O.)
5. INIAV, INIAV-Dois Portos, Quinta da Almoínha, 2565-191 Dois Portos, Portugal; ilda.caldeira@iniav.pt (I.C.); sara.canas@iniav.pt (S.C.)
6. MED—MediterraneanInstitute for Agriculture, Environment and Development, Instituto de formação avançada, Universidade de Évora, Pólo da Mitra, Ap. 94, 7006-554 Évora, Portugal
* Correspondence: ofelia@ipcb.pt; Tel.: +351-272-339-900

Received: 28 April 2020; Accepted: 27 May 2020; Published: 2 June 2020

Abstract: Fourier transform infrared spectroscopy (FTIR) with Attenuated Total Reflection (ATR) combined with functional data analysis (FDA) was applied to differentiate aged wine spirits according to the ageing technology (traditional using 250 L wooden barrels versus alternative using micro-oxygenation and wood staves applied in 1000 L stainless steel tanks), the wood species used (chestnut and oak), and the ageing time (6, 12, and 18 months). For this purpose, several features of the wine spirits were examined: chromatic characteristics resulting from the CIELab method, total phenolic index, concentrations of furfural, ellagic acid, vanillin, and coniferaldehyde, and total content of low molecular weight phenolic compounds determined by HPLC. FDA applied to spectral data highlighted the differentiation between all groups of samples, confirming the differentiation observed with the analytical parameters measured. All samples in the test set were differentiated and correctly assigned to the aged wine spirits by FDA. The FTIR-ATR spectroscopy combined with FDA is a powerful methodology to discriminate wine spirits resulting from different ageing technologies.

Keywords: FTIR-ATR; FDA; vector analysis; wine spirit; ageing technology; micro-oxygenation; wood; oak; chestnut; ageing time

1. Introduction

The contact of wine distillate with wood is recognised as a pivotal step of the aged wine spirit production, during which its quality increases and sensory fullness can be reached. Scientific evidence exists on the key role of several physicochemical phenomena, particularly the extraction and oxidation reactions involving the wood-derived compounds of low molecular weight, on the chemical changes (quantitative and qualitative aspects of the beverage's volatile and non-volatile composition) and sensory changes (colour, aroma, and taste) occurred [1–3]. Besides, they mainly depend on the ageing technology, the kind of wood used (oak and chestnut), and the length of the ageing process [1,4,5].

Despite the high-quality spirits attained by the traditional ageing technology, using wooden barrels, this expensive and lengthy process led to the search and study of sustainable alternative technologies. Research has been focused on a technology based on adding wood pieces to the distillate kept in stainless steel tanks, as for other alcoholic beverages [6,7]. Recently, the micro-oxygenation technique, reproducing the oxygen transfer that occurs in the wooden barrel, was applied to optimise this ageing technology. Promising outcomes based on the phenolic composition and colour of the wine spirits in the beginning of ageing were attained [8]. Nevertheless, a comprehensive approach is needed towards the full/robust characterisation of this novel technology, exploring the data acquired over the ageing period through different analytical and statistical methodologies.

Spectroscopic techniques are very useful for food and beverages quality evaluation because they require minimal or no sample preparation (absence of extraction, reactions with some other chemical species, treatment with a chelating agent, masking, sub-sampling, or other manipulation), they are rapid and non-destructive (causing no physicochemical changes during the analysis), and they can be used to simultaneously assess several parameters of a sample. In recent years, significant developments related to the applicability of vibrational spectroscopy combined with multivariate data analysis has been made to give the rapid quantification of several compounds in different matrices or to discriminate different groups of samples. Concerning the alcoholic beverages, the studies were more centred on other drinks than wine spirits. FTIR-ATR (Fourier transform infrared spectroscopy with Attenuated Total Reflection) was used to evaluate different parameter in alcoholic beverages, such as the determination of important quality parameters of beers [9], determination of ethanol content in liquors [10], quantitative analysis of methanol (an adulterant in alcoholic beverages) [11], analysis of ethanol and methanol content in traditional fruit spirits [12], and the authentication of whisky according to its botanical origin [13]. Actually, the studies on wine spirits with FTIR-ATR are scarce. One of them was carried out by Anjos et al. [14], applying the FTIR-ATR methodology to predict the alcoholic strength, methanol, acetaldehyde, and fusel alcohols contents of grape-derived spirits.

In this paper, a new application of functional data analysis (FDA) in the field of quality evaluation using spectroscopic techniques is presented. In recent years, FDA has been used in numerous investigations to analyse processes in continuous time; some examples are energy efficiency [15], medical research [16], econometrics [17], optimisation problems [18], industrial processes [19], environmental research [20,21], and food science [22]. In all these works, FDA showed its usefulness for the study of functions, defined in a specific interval, without missing the correlation between the observations.

FDA allows the analysis of the entire curves, which represent individual observations of the sample under study, detecting different behaviours throughout the curves [15,23]. Ordoñez et al. [23], in a similar context that this paper, showed significant reasons to analyse the sample with curves instead of individual observations. In addition, the contrast of similarity has been carried out from a vectorial approach, but summarising the curves with a single value; in this case, the mean. This is necessary because the curves, although representing individual data, are formed by a set of observations correlated to each other. Martínez et al. [20] explained how this correlation is not taken into account from a vectorial approach. Furthermore, from this point of view, to contrast the similarity between the samples, a classical analysis of variance (ANOVA) [24,25] and anon-parametric Kruskal's test [26] have been performed. Although this is a simpler approach, it contributes to highlight the strengths of FDA. On the one hand, Martínez et al. [15] presented how different conclusions can be reached for each approach because of the biased sample used in the vector analysis (mean of the curves) or different detection of outliers. On the other hand, among others, Sancho et al. [19] demonstrated that FDA presents more realistic and accurate results, showing how and at which specific part of the curve the groups have different spectrometric behaviours.

The aim of this research is to contrast the similarity between the samples obtained with FTIR-ATR spectrometry for wine spirits aged by different ageing technologies (traditional and alternative), with different kinds of wood (chestnut and oak) and over the ageing period (6, 12, and 18 months), using functional data analysis of variance (FANOVA).

Thus, the second section presents, on the one hand, the data used in this study and, on the other hand, the specific methodology applied to obtain the results. The third section presents the results of the comparison and analysis of the differences between the wine spirit samples. Then, in the fourth section, a discussion is carried out about the results and the information obtained with them. Finally, in Section 5, the conclusions of this work are presented.

2. Materials and Methods

2.1. Samples

A total of 30 samples of wine spirit (produced at industrial scale) aged with Limousinoak (*Quercus robur* L.) and Portuguese chestnut (*Castanea sativa* Mill.) for 6, 12, and 18 months of ageing time were analysed (Figure 1), covering the following categories:

- TL—aged in 250 L new oak wooden barrels;
- TC—aged in 250 L new chestnut wooden barrels;
- AL—aged in 1000 L stainless steel tanks with oak wood staves and micro-oxygenation;
- AC—aged in 1000 L stainless steel tanks with chestnut wood staves and micro-oxygenation.

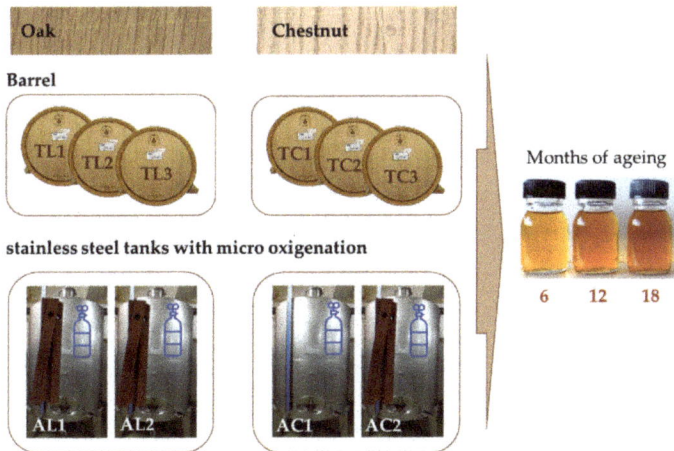

Figure 1. Scheme of the essay. 1, 2, and 3 indicate the essay replicates.

From each sample, four independent samples were taken for spectral analysis, totalising 120 spectra analysed.

The wooden barrels and the stainless-steel tanks were placed in the cellar in similar environmental conditions. The same wine distillate (alcohol strength, 77.4 *v/v*; pH, 5.44; total acidity, as acetic acid, 0.13 g/hL of absolute ethanol; volatile acidity, as acetic acid, 0.11 g/hL of absolute ethanol) produced by the Adega Cooperativa da Lourinhã, Portugal, was used. The wood pieces and barrels were manufactured by J.M. Gonçalves cooperage (Palaçoulo, Portugal) with the medium plus toasting level, as described by Canas et al. [8].

To characterise the wine spirits aged by different technologies, with different wood species, through the ageing time, several analytical determinations were performed for all the samples: CIELAB colour parameters (lightness, saturation, and chromaticity coordinates), total phenolic index, and low molecular weight compounds contents, according to the methodologies described below. All the analyses were done in duplicate.

2.2. Analytical Procedures

2.2.1. Analysis of Chromatic Characteristics

The chromatic characteristics of the wine spirits—lightness (L *), varying between 100% (fully transparent) and 0% (fully opaque); saturation (C *); chromaticity coordinates (a * and b *), of which the coordinate a * varies between green (a * < 0) and red hues (a * > 0), and the coordinate b * varies between blue (b * < 0) and yellow hues (b * > 0)—were analysed according to the CIELab method. The analysis was performed in a Varian Cary 100 Bio spectrophotometer (Santa Clara, CA, USA) with a 10 mm glass cell, considering a D65 illuminant and a 10° standard observer. The transmittance measurement was made every 10 nm from 380 to 770 nm. The analysis was performed in duplicate.

2.2.2. Determination of the Total Phenolic Index

The total phenolic index (TPI) of the wine spirits was analysed as described by Cetó et al. [27]: dilution of the samples with ethanol/water 77:23 v/v; absorbance measurement at 280 nm, using a Varian Cary 100 Bio spectrophotometer (Santa Clara, CA, USA) with a 10 mm quartz cell; calculation of the total phenolic index by multiplying the measured absorbance by the dilution factor. The analysis was performed in duplicate.

2.2.3. Analysis of Low Molecular Weight Compounds

The phenolic and furanic compounds, namely phenolic acids (gallic acid, vanillic acid, syringic acid and ellagic acid), phenolic aldehydes (vanillin, syringaldehyde, coniferaldehyde, and sinapaldehyde), coumarins (umbelliferone and scopoletin), and furanic aldehydes (5-hydroxymethylfurfural, 5-methylfurfural, and furfural) were quantified by liquid chromatography according to the method proposed by Canas et al. [28]. All compounds were quantified to calculate the total content of low molecular weight compounds. Only some compounds—furfural, ellagic acid, vanillin, and coniferaldehyde—were considered in the discussion of individual compounds contents based on their highly discriminant power.

The chromatographic analysis was carried out in a HPLC Lachrom Merck Hitachi system (Merck, Darmstadt, Germany) composed of a quaternary pump L-7100, a column oven L-7350 equipped with a 250 mm × 4 mm i.d. LiChrospher RP 18 (5 μm) column (Merck, Darmstadt, Germany), a UV–Vis detector L-7400, a fluorescence detector L-7480 (connected to the UV–Vis detector), and an autosampler L-7250. The HSM D-7000 software (Merck, Darmstadt, Germany) was used for the management, acquisition, and treatment of the data. The following chromatographic conditions were used: (i) binary gradient consisting of solvent A, water: formic acid (98:2 v/v) and solvent B, methanol:water:formic acid (70:28:2 $v/v/v$) as follows: 0% isocratic B in 3 min, linear gradient from 0% to 40% B in 22 min, from 40% to 60% B in 18 min, 60% isocratic B in 12 min, linear gradient from 60% to 80% B in 5 min, 80% isocratic B in 5 min; (ii) column temperature of 40 °C; (iii) flow rate of 1 mL/min. Phenolic acids and furanic aldehydes were detected at 280 nm, phenolic aldehydes were detected at 320 nm, and coumarins were detected at 325 nm (excitation)/454 nm (emission).

Samples were added with an internal standard (20 mg/L of 4-hydroxybenzaldehyde), filtered through 0.45 μm membrane (Titan, Scientific Resources Ltd., Gloucester, UK) and analysed by the direct injection of 20 μL. The identification of chromatographic peaks was made through comparison of their retention time and UV-Vis spectra with those commercial standards. The assessment of chromatographic purity of the peaks and their UV–Vis spectra (200–400 nm) were performed in Waters system composed of a photodiode-array detector (Waters 996), in the same chromatographic conditions and managed by 'Millennium 2010' software (Waters, Milford, NA, USA). The analysis was performed in duplicate. The quantification of each compound was based on a calibration curve made with the corresponding commercial standard.

2.2.4. Spectroscopic Analyses

Spectra were acquired by the Fourier transform infrared spectroscopic method with platinum Attenuated Total Reflectance (FTIR-ATR) with a Bruker spectrometer (Alpha, Bruker Optic GmbH, Ettlingen, Germany) using a diamond crystal. Four spectra per sample were obtained with 128 scans per spectrum at a spectral resolution of 8 cm^{-1} in the range of 4000 to 450 cm^{-1}.

The FTIR-ATR used was equipped with a flow-through cell with controlled temperature. The cleaning of the cell was done by the injection of water in the flow-through cell, and the background was also measured with distilled water.

2.3. Statistical Analysis

2.3.1. Statistical Treatment of Analytical Data

A two-way analysis of variance (ANOVA) was carried out to examine the influence of the two fixed factors—ageing technology (two levels: alternative versus traditional) and kind of wood (two levels: Limousin oak versus chestnut)—on the chromatic characteristics, total phenolic index, and low molecular weight compounds' contents of the wine spirits after 6, 12, and 18 months of ageing. For each significant factor or interaction, the variance percentage was calculated. Fisher's least significant difference (LSD) test was used to compare the average values observed for each sample group. All the calculations were carried out using Statistica vs. 5 (Statsoft Inc., Tulsa, OK, USA).

2.3.2. Functional Analysis

The similarity between the different samples of ageing technology and the different samples of ageing time was contrasted. This analysis was carried out from a vectorial and functional approach. Vectorially, the tests used were the classical ANOVA [25], comparing the mean levels of the groups, and Kruskal's non-parametric test [26], which studies whether the observations of each group come from the same distribution. In addition, from the functional approach, the functional ANOVA (FANOVA) was performed.

Functional Data Analysis (FDA)

The analyses from a functional approach study functions, based on sets of observations, were defined in a determined interval I = [a,b]. One of its strengths is its structure of infinite dimensions that allows to extend the possibilities of data analysis [29,30]. A random variable, measured at a set of discrete points $\{t_g\}_{g=1}^{G} \in [a,b]$, has to take values in metric or semi-metric spaces to be considered functional. Thus, functional data take the form of a matrix with n rows, one for each individual studied, and G columns representing the points of evaluation of the functions [31,32].

Smoothing is the most used process to convert discrete observations into continuous functions, $x(t)$, $t \in X \subset \mathcal{F}$; where $\mathcal{F}$ is the functional space. Specifically, assuming that the functions are observed with error, a functional basis expansion can be adopted as follows:

$$x(t) = \sum_{w=1}^{W} c_w \phi_w(t) \tag{1}$$

where $\{c_w\}_{w=1}^{W}$ is the w-th basis coefficients, $\{\phi_w(t)\}_{w=1}^{W}$ is the w-th basis function, and W is the number of basis functions under consideration [29,32]. Thus, the basis functions used in this work were splines [33] due to their specific properties such as the possibility of generating large basis sets easily or their flexibility [34]. On the other hand, to select the number of bases of each sample, the determination coefficient R^2 was taken into account. The number of bases is the minimum number at which R^2

stops improving significantly or surpasses the value of 0.99 (see Martínez et al. [15]). Moreover, the smoothing process involves solving the following problem:

$$\min_{x \in F} \sum_{g=1}^{G} \{z_g - x(t_g)\}^2 + \lambda \Gamma(x) \qquad (2)$$

where $z_g = x(t_g) + \epsilon_g$ is the value obtained by evaluating x at point t_g with ϵ_g being a random noise with zero mean, λ is a parameter controlling the intensity of regularisations, and Γ is a parameter that makes it costly to reach complex solutions. Then, the basis coefficients can be expressed as the solution of the smoothing process as follows [29,35]:

$$c = (\Phi^t \Phi + \lambda R)^{-1} \Phi^T z \qquad (3)$$

being Φ a GxW matrix formed by $\Phi_{gw} = \phi_w(t_g)$ and R being a WxW matrix of the elements $R_{wg} = \int_T D^2 \phi_w(t) D^2 \phi_g(t) dt$ where $D^n \phi_w(t)$ is the nth-order differential operator of ϕ_w.

Functional Depths

The depth concept, in classical multivariate statistics, was used to measure the centrality of a point $x \in \mathbb{R}^d$ within a data set. The points nearest to the centre obtain a higher depth value [36]. With a functional approach, depths measure the centrality of a curve in relation to the other curves of the sample $x_1, \ldots, x_n$, coming from a stochastic process $\mathcal{X}(\cdot)$ evaluated at a specific interval $[a, b] \in \mathbb{R}$ [37,38].

Although there exist different functional types of depths (Fraiman-Muniz [37], h-modal [39] or Random Projections [38]), the most widely used is the h-modal depth due to its better performance in the correct detection of outliers [36]. Therefore, the functional mode of the sample will be the curve most densely surrounded by other curves. The functional depth of a certain curve in a specific sample is calculated as follows:

$$MD_n(x_i, h) = \sum_{w=1}^{n} K\left(\frac{\|x_i - x_w\|}{h}\right) \qquad (4)$$

where $\|\cdot\|$ is the norm in a functional space, $K : \mathbb{R}^+ \to \mathbb{R}^+$ is a kernel function, and h represents the bandwidth parameter [39].

Functional depths, which are considered as a measure of dispersion, are essential in the detection of outliers. In any data analysis, the identification of these atypical data is crucial because they could affect the subsequent estimations. In addition, examining them is important to discover the causes that give these observations a different behaviour from the rest. Besides, in functional analyses, it is even more important because it is possible that individually, the values that form the curve are not outliers in a vectorial way but, from a functional point of view, the entire curve could be [36,40]. Martínez et al. [29] explained in detail how to detect functional outliers within a functional sample.

Functional ANOVA (FANOVA)

Functional data Analysis of Variance, similar to the vector version, contrasts the distance between the mean levels of the factor variables. The aim of this contrast is to find out if the set of functions studied are statistically distinguishable [41]. There will also be Q independent samples $X_{gj}(t)$, $j = 1, \ldots, n_g$; $t \in I = [a, b]$. But these samples are extracted from $\mathcal{L}^2(I)$ processes $X_g(t)$, $g = 1, \ldots, Q$ and their mean function is $E(X_g(t)) = m_g(t)$ [42,43]. If the functional sample is divided into groups like $\{X_j, \mathcal{A}_j\}_{j=1}^n \in \mathcal{F}$ x$\mathcal{A} = \{1, \ldots, A\}$, being $\mathcal{A}$ the factor variable, the hypothesis contrast has the following form:

$$\begin{cases} H_0 : \overline{X}_1(t) = \overline{X}_2(t) = \ldots = \overline{X}_A(t) \\ H_1 : \exists h, e \text{ s.t. } \overline{X}_h(t) \neq \overline{X}_e(t) \end{cases} \qquad (5)$$

The model for the j-th observation belonging to the g-th group has the following form [41]:

$$X_{jg}(t) = \overline{X}(t) + \alpha_g(t) + \epsilon_{jg}(t) \qquad (6)$$

where $X_{jg}(t)$ is the functional value of group g, $\alpha_g(t)$ is the effect of being part of a determined group and $\epsilon_{jg}(t)$ represents the unexplained variability for the i-th observation of group g. Furthermore, the model in Equation (6) can be represented in its matrix notation:

$$\mathbf{X}(t) = \mathbf{Z}\boldsymbol{\gamma}(t) + \boldsymbol{\epsilon}(t) \qquad (7)$$

being $\mathbf{X}(t)$ a N-dimensional vector, $\boldsymbol{\gamma}(t) = (\overline{X}(t), \alpha_1(t), \ldots, \alpha_Q(t))^T$ a (Q + 1)-dimensional vector, $\boldsymbol{\epsilon}(t)$ a vector of N residual functions, and $\mathbf{Z}$ the design matrix with dimension (N, Q + 1).

Thus, to assure the indentification of the functional effects $\alpha_g(t)$, the sum to zero constraint is introduced [41,44]:

$$\sum_{g=2}^{Q+1} \boldsymbol{\gamma}(t) = 0, \ \forall t \qquad (8)$$

The parameter vector $\boldsymbol{\gamma}(t)$ in Equation (7) can be estimated minimising the standard least squares:

$$LMSSE(\boldsymbol{\gamma}) = \int [\mathbf{X}(t) - \mathbf{Z}\boldsymbol{\gamma}(t)]^T [\mathbf{X}(t) - \mathbf{Z}\boldsymbol{\gamma}(t)] dt$$

subject to the constraint (8) [41,44].

Regarding the contrast in Equation (5), most tests are based on F test statistic [32,45]:

$$F_n(t) = \frac{SSR_n(t)/(Q-1)}{SSE_n(t)/(n-Q)}$$

where

$$SSR_n(t) = \sum_{g=1}^{Q} n_g (\overline{X}_g(t) - \overline{X}(t))^2$$

$$SSE_n(t) = \sum_{g=1}^{Q} \sum_{j=1}^{n_g} (X_{gj}(t) - \overline{X}_g(t))^2$$

represents the variations between groups and within groups, respectively. In addition, for these calculations, the sample mean function $\overline{X}(t) = (1/n) \sum_{g=1}^{Q} \sum_{j=1}^{n_g} X_{gj}(t)$ and the sample group mean function $\overline{X}_g(t) = (1/n_g) \sum_{j=1}^{n_g} X_{gj}(t)$, $g = 1, \ldots, Q$ were taken into account.

In this work, two specific tests were used to contrast the similarity between samples. On the one hand, the F test with the reduced bias estimation method (FB) [46]. This test uses both point variations between groups and variations within groups. Specifically, the test statistic has the form:

$$F_n = \frac{\int_Q SSR_n(t)dt/(Q-1)}{\int_Q SSE_n(t)/(n-Q)} \qquad (9)$$

The distribution of this statistic is approximated by $F_{(Q-1)k,(n-Q)k}$, where k is estimated by the bias-reduced method. The p-value taken into account comes from $P(F_{(Q-1)k,(n-Q)k} > F_n)$ [45,46].

On other hand, a permutation test based on a representation of the base function (FP) was used. This test is based on the basis representation procedure presented by Górecki and Smaga [47].

The functional observations are be represented by a finite number of basis functions $\varphi_m \in \mathcal{L}^2(I)$, $m = 1, \ldots, K$ as follows:

$$X_{gj}(t) \approx \sum_{m=1}^{K} c_{gjm}\varphi_m(t), \ t \in I \tag{10}$$

where c_{gjm} are random variables with a significantly large K. Moreover, this test uses the following approximation of Equation (9) [45]:

$$\frac{(a-b)/(Q-1)}{(c-a)/(n-Q)} \tag{11}$$

where

$$a = \sum_{g=1}^{Q} \frac{1}{n_g} 1_{n_g}^T \mathbf{C}_g^T \mathbf{J}_\varphi \mathbf{C}_g 1_{n_g}$$

$$b = \frac{1}{n}\sum_{g=1}^{Q}\sum_{j=1}^{Q} 1_{n_g}^T \mathbf{C}_g^T \mathbf{J}_\varphi \mathbf{C}_j 1_{n_j}$$

$$c = \sum_{g=1}^{Q} trace(\mathbf{C}_g^T \mathbf{J}_\varphi \mathbf{C}_g)$$

being 1_a a vector of ones with dimension $ax1$, $\mathbf{C}_g = (c_{gjm})_{j=1,\ldots,n_g;m=1,\ldots,K}$, $i = 1,\ldots,Q$, and $\mathbf{J}_\varphi := \int_I \boldsymbol{\varphi}(t)\boldsymbol{\varphi}^T(t)dt$ is the matrix of cross-products with dimensions KxK based on $\boldsymbol{\varphi}(t) = (\varphi_1(t),\ldots,\varphi_K(t))^T$.

3. Results

The two-way ANOVA results (Table 1) show that the ageing technology and the kind of wood had a highly significant effect on the colour and total phenolic content acquired by the wine spirits during the ageing process (after 6, 12, and 18 months). Among these factors, greater influence was exerted by the ageing technology (higher percentage of the variance explained) on the chromatic characteristics, while a similar weight of both factors was observed in the phenolic content. Regardless of the sampling time, the wine spirits aged with micro-oxygenation and chestnut wood staves (AC) exhibited a significantly lower value of lightness (L *) and significantly higher values of saturation (C *) and chromaticity coordinates (a * and b *) than the others. This set of chromatic characteristics reflects a more evolved colour, since lower L * and higher C * correspond to a more intense/darker colour, and the combination of higher a * and b * (yellow and red hues, respectively) is associated with a greater intensity of amber or orange hue, which made these spirits look older than the others. The colour of wine spirits from Limousin wooden barrels (TL) was on the opposite side, while the wine spirits aged with micro-oxygenation and Limousin wood staves (AL) and those aged in chestnut wooden barrels (TC) presented intermediate characteristics.

Table 1. Effect of the ageing technology and kind of wood on the chromatic characteristics and total phenolic index acquired by the wine spirits after 6, 12, and 18 months of ageing. AC: Alternative Chestnut, AL: Alternative Oak, TC: Traditional Chestnut, TL: Traditional Oak.

Ageing Months		Code	L *(%)	A *	B *	C *	TPI
6		TC	85.41 ± 1.41 [b]	3.37 ± 1.08 [b]	50.96 ± 2.68 [b]	51.08 ± 2.74 [b]	24.94 ± 1.98 [b]
		TL	93.73 ± 0.42 [c]	−1.25 ± 0.17 [a]	26.76 ± 2.21 [a]	26.79 ± 2.20 [a]	11.99 ± 1.18 [a]
		AC	77.14 ± 1.26 [a]	11.79 ± 1.22 [c]	70.00 ± 1.59 [c]	70.99 ± 1.77 [c]	47.79 ± 3.80 [c]
		AL	87.55 ± 0.23 [b]	1.75 ± 0.14 [b]	46.24 ± 0.14 [b]	46.27 ± 0.15 [b]	26.88 ± 0.87 [b]
Variance origin		Technology (S)	61.9 ***	52.7 ***	60.2 ***	60.2 ***	42.1 ***
		Wood (W)	36.7 ***	31.9 ***	38.8 ***	38.8 ***	52.4 ***
		SxW	NS	14.0 **	NS	NS	4.2 *
		Residual	1.5	1.3	1.0	1.0	1.3
12		TC	79.11 ± 1.38 [b]	9.89 ± 1.31 [b]	69.41 ± 1.64 [b]	70.11 ± 1.80 [b]	37.90 ± 1.94 [b]
		TL	90.62 ± 1.04 [c]	−0.49 ± 0.65 [a]	40.66 ± 3.4 [a]	40.67 ± 3.40 [a]	17.55 ± 1.81 [a]
		AC	65.58 ± 1.97 [a]	25.63 ± 1.78 [c]	87.25 ± 0.56 [c]	90.95 ± 1.04 [c]	65.72 ± 1.37 [c]
		AL	79.17 ± 0.60 [b]	10.38 ± 0.53 [b]	70.87 ± 0.61 [b]	71.63 ± 0.68 [b]	37.86 ± 0.23 [b]
Variance origin		Technology (S)	49.2 ***	49.9 ***	49.4 ***	50.7 ***	48.6 ***
		Wood (W)	49.7 ***	46.3 ***	43.5 ***	44.9 ***	48.8 ***
		SxW	NS	3.1 *	6.2 ***	3.6 *	2.2 *
		Residual	1.1	0.8	0.9	0.8	0.5
18		TC	77.33 ± 1.25 [b]	12.06 ± 1.24 [b]	73.97 ± 1.22 [b]	74.95 ± 1.40 [b]	40.98 ± 2.46 [b]
		TL	89.61 ± 1.17 [c]	−0.02 ± 0.84 [a]	44.35 ± 3.74 [a]	44.36 ± 3.74 [a]	18.15 ± 1.63 [a]
		AC	62.29 ± 1.94 [a]	28.97 ± 1.62 [c]	89.27 ± 0.12 [c]	93.86 ± 0.61 [c]	71.60 ± 2.50 [c]
		AL	76.59 ± 0.52 [b]	13.19 ± 0.45 [b]	75.87 ± 0.44 [b]	77.01 ± 0.51 [b]	40.55 ± 1.25 [b]
Variance origin		Technology (S)	52.2 ***	52.9***	47.7 ***	50.0 ***	47.9 ***
		Wood (W)	46.9 ***	45.2***	40.2 ***	42.3 ***	49.5 ***
		SxW	NS	1.4*	11.1 **	6.8 ***	2.1 *
		Residual	0.9	0.6	0.9	0.9	0.6

L *—lightness; a *, b *—chromaticity coordinates; C *—saturation; TPI—total phenolic index. For each sampling time (6, 12 and 18 months), mean values with the same letter in a column are not statistically different. NS, $p > 0.05$; * $0.01 < p < 0.05$; ** $0.001 < p < 0.01$; *** $p < 0.001$.

In a previous work, this research team have already identified furfural, ellagic acid, vanillin and coniferaldehyde as markers of the ageing technology and the kind of wood used [8]. The results obtained for these compounds are shown in Table 2. It should be stressed that most of the low molecular weight compounds contents were mainly dependent on the ageing technology (higher percentage of the variance explained), as aforementioned for the chromatic characteristics and TPI, but ellagic acid content was closely related to the kind of wood. Indeed, significantly higher levels of furfural, vanillin, and coniferaldehyde were found in the wine spirits aged by the alternative technology; the first two were higher in the modality comprising chestnut wood staves (AC), whereas the latter was higher in the modality comprising Limousin wood staves (AL). Slight differences were found between the wine spirits from chestnut barrels (TC) and Limousin oak barrels (TL). Regarding ellagic acid, higher content was promoted by the chestnut wood (AC and TC), especially in the alternative technology. The contact with Limousin oak together with micro-oxygenation (AL) induced a slightly lower content of this phenolic acid, and a weak performance was showed by the Limousin oak barrels (TC).

Besides the differences between the two ageing technologies and the two kinds of wood, the results also reveal a remarkable role of the ageing time, as indicated by the percentage of variance observed for each factor in ANOVA (Table 2). Regardless of the ageing modality, there was a gradual decrease of lightness and a gradual increase of saturation and chromatic coordinates that correspond to the colour development over the ageing process.

Table 2. Effect of the ageing technology and kind of wood on the contents of low molecular weight compounds (mg/L absolute ethanol) of the wine spirits after 6, 12, and 18 months of ageing.

Ageing Months	Code	Furfural	Ellagic Acid	Vanillin	Coniferaldehyde	sumHPLC
6	TC	38.31 ± 6.90 [a]	8.12 ± 1.41 [b]	2.03 ± 0.01 [b]	6.17 ± 0.63 [a]	163.10 ± 24.08 [b]
	TL	31.73 ± 6.38 [a]	3.43 ± 0.20 [a]	1.49 ± 0.08 [a]	5.21 ± 0.04 [a]	78.99 ± 9.71 [a]
	AC	127.05 ± 5.07 [c]	15.35 ± 0.38 [c]	4.62 ± 0.20 [d]	10.60 ± 0.65 [b]	296.05 ± 15.91 [c]
	AL	87.74 ± 4.11 [b]	6.28 ± 0.70 [a,b]	3.26 ± 0.16 [c]	12.20 ± 0.52 [b]	195.23 ± 3.04 [b]
Variance origin	Technology(S)	82.7 ***	30.4 ***	79.1 ***	98.5 ***	63.1 ***
	Wood(W)	8.1 **	56.9 ***	15.1 ***	NS	34.6 ***
	SxW	8.0 ***	10.8 ***	5.3 **	NS	NS
	Residual	1.2	2.0	0.5	1.5	2.3
12	TC	35.85 ± 6.03 [a]	12.86 ± 1.16 [b]	3.61 ± 0.22 [b]	7.00 ± 0.57 [a]	231.68 ± 26.34 [b]
	TL	31.36 ± 5.80 [a]	5.64 ± 0.34 [a]	2.66 ± 0.23 [a]	6.49 ± 0.58 [a]	95.09 ± 13.67 [a]
	AC	131.17 ± 4.91 [c]	24.89 ± 1.26 [c]	8.68 ± 0.02 [d]	13.97 ± 0.17 [b]	369.24 ± 8.57 [d]
	AL	96.08 ± 1.93 [b]	11.94 ± 0.88 [b]	6.77 ± 0.09 [c]	19.61 ± 0.41 [c]	275.32 ± 4.56 [c]
Variance origin	Technology(S)	87.9 ***	41.3 ***	89.2 ***	79.7 ***	64.5 ***
	Wood(W)	5.2 **	50.0 ***	8.6 ***	5.1 ***	33.8 ***
	SxW	6.1 ***	7.8 ***	1.8 ***	14.8 ***	NS
	Residual	0.8	0.9	0.3	0.4	1.7
18	TC	36.55 ± 7.28 [a]	15.48 ± 1.41 [b]	4.43 ± 0.34 [b]	6.85 ± 0.75 [a]	271.85 ± 38.84 [b]
	TL	31.63 ± 6.26 [a]	6.81 ± 0.49 [a]	3.13 ± 0.23 [a]	6.41 ± 0.60 [a]	104.10 ± 16.10 [a]
	AC	113.35 ± 4.27 [c]	28.17 ± 1.15 [c]	8.61 ± 0.07 [d]	11.16 ± 0.15 [b]	347.46 ± 4.33 [c]
	AL	86.72 ± 0.21 [b]	13.81 ± 0.23 [b]	7.49 ± 0.24 [c]	17.93 ± 0.07 [c]	275.34 ± 4.79 [b]
Variance origin	Technology(S)	89.4 ***	39.1 ***	92.0 ***	63.3 ***	43.6 ***
	Wood(W)	4.8 ***	53.7 ***	7.2 ***	10.0 ***	40.9 ***
	SxW	4.3 *	6.2 ***	NS	26.0 ***	11.9 *
	Residual	1.5	0.8	0.7	0.7	3.6

Furf—furfural; Ellag—ellagic acid; Vanil—vanillin; Cofde—coniferaldehyde; sumHPLC—total content of low molecular weight compounds determined by HPLC. For each sampling time (6, 12, and 18 months), mean values with the same letter in a column are not. NS, $p > 0.05$; * $0.01 < p < 0.05$; ** $0.001 < p < 0.01$; *** $p < 0.001$.

The levels of significance shown in Table 3 reveal that the variation of chromatic characteristics, phenolic content (TPI and sumHPLC), and most of individual compounds contents were mostly significant between 6 months and 12 months. Significant differences between the three sampling times marked the wine spirits aged by the alternative technology with Limousin staves. In general, the L * parameter decreased over the ageing time, while the other parameters (a *, b *, and C *) increased. Furfural, ellagic acid, vanillin, coniferaldehyde and total content of low molecular weight compounds increased. Similar results were observed in the first months of ageing time of this kind of wine spirit [8].

Table 3. Differences observed for each samples group (technology vs. wood) during the ageing time.

	TC			TL			AC			AL		
	6	12	18	6	12	18	6	12	18	6	12	18
L *(%)	b	a	a	b	a	a	b	a	a	c	b	a
a *	a	b	b	a	a	a	a	b	b	a	b	c
b *	a	b	c	a	b	b	a	b	b	a	b	c
C *	a	b	c	a	b	b	a	b	b	a	b	c
TPI	a	b	b	a	b	b	a	b	b	a	b	b
Furf	a	a	a	a	a	a	a	ab	b	a	ab	b
Ellag	a	b	b	a	b	c	a	b	b	a	b	c
Vanil	a	b	c	a	b	c	a	b	b	a	b	c
Cofde	a	a	a	a	b	b	a	b	a	a	b	c
sumHPLC	a	b	b	a	a	a	a	b	b	a	b	b

The spectra obtained for wine spirit are similar to those reported by others authors [28].

The representative absorbance spectra of wine spirit samples studied is plotted in Figure 2; their spectral information is in accordance with previous reports of other authors [48]. The IR region from 2990 to 3626 cm^{-1} has a very strong influence due to water present in the samples [49]. Nevertheless, for these analyses, the background was measured with water.

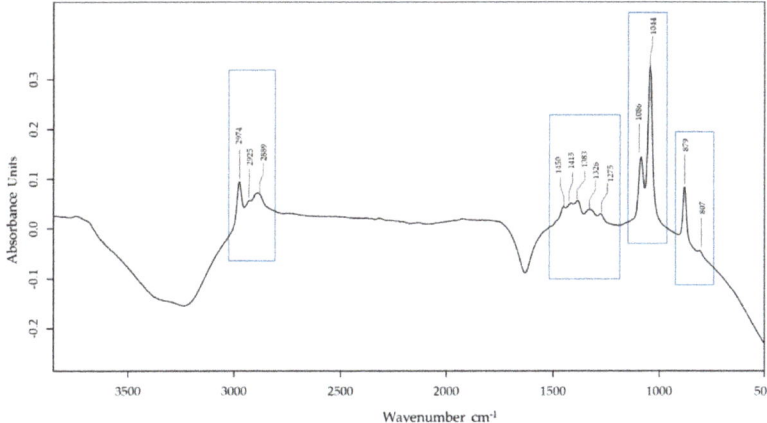

Figure 2. Fourier transform infrared spectroscopy–Attenuated Total Reflection (FTIR-ATR) absorbance spectra of wine spirit samples.

There is IR information in the regions from 3000 to 2900 cm^{-1} due to the O–H stretching of alcohols and C–H stretching of CH_3 and CH_2, and consequently related to the presence of ethanol and methanol in the alcoholic beverages [49].

Regarding the region from 1500 to 860 cm^{-1}, it corresponds to C–C and C–O vibrations in volatile compounds [12,14].

The small peak at 1450 cm^{-1} was assigned to C–OH bending deformation, and the peak at 1275 was assigned to C–O stretching in the acid molecules [11,50].

According to Stuart (2004), the region from 1300 to 840 cm^{-1} shows other absorption bands assigned to the C=O and C=C groups present in furanic compounds. The highest peaks at 1086 and 1044 cm^{-1} are ascribed to the C–O stretch absorption bands, which are important regions for ethanol and methanol identification and quantification respectively, and C–C absorption bands, which are related to ethanol and some organic compounds such as sugars, phenols, alcohols, and esters [14,49,51].

The peak at 879 cm^{-1} could be ascribed to out-of-plane C–H bending of aromatic compounds [10], and to CH–OH, C–C, C–O, and C–H bond stretching due to water, sugars, and phenolic compounds [51].

According to these regions of the FTIR-ATR absorption spectra (with baseline correction) of the wine spirit, a mathematical analysis has been performed to determine the differences between the groups studied. On the one hand, a functional ANOVA (FANOVA) model has been considered using two different tests: FP, the permutation test based on a representation of the base function (Equation (9)); and FB, the F test with the reduced bias estimation method (Equation (11)). On the other hand, a vector analysis based on the classical Analysis of Variance and the non-parametric Kruskal test have been carried out. Although all the areas of the curves have been analysed following the same process, only two were plotted each time to show the differences between the samples (space problem).

First, the different groups based on the ageing technology and kind of wood (AC, AL, TC, and TL) were tested. The contrast is different for each ageing time (18 months, 12 months, and 6 months). With an 18-month period, the hypothesis of similarity between all groups is rejected in all areas of the whole curve and from both points of view (Table 4). The samples of wine spirits aged by the alternative technology, on average, have higher absorbance units than the aged by the traditional one. Within the alternative technology, the wine spirit aged in oak wood always showed higher absorbance units than the one aged in chestnut wood (Figure 3). With an ageing period of 12 months, the similarity hypothesis is also rejected in all areas by the two analyses (Table 4). Even so, in this case, the differences between alternative and traditional ageing technologies are small. The spectrometric curves can hardly be differentiated in the functional part of Figure 4, and the *p*-values obtained are higher (Table 4). With an ageing time of 6 months, significant reasons were found to reject the similarity hypothesis in all areas through functional and vector analysis (Table 4). With this sample, the difference is more significant than with the 12-month sample but less than with the 18-month sample. Moreover, it can be seen that the wine spirit that gets higher absorbance units in the alternative sample is the one aged in oak, but with the traditional sample, it is the one aged in chestnut (Figure 5). These results are in accordance with those previously observed for the colour and analytical parameters (Tables 1 and 2).

Table 4. Numerical results of the similarity contrast between the groups AC, AL, TC, and TL, depending on the ageing time of the wine spirit samples. Functional results (FDA) are based on FANOVA using two different tests (FP: permutation test based on a representation of the base function, FB: F test with a reduced bias estimation method). In addition, the results of the ANOVA and Kruskal's test representing the vectorial results (VA) are shown. All the results are *p*-values based on a 5% significance level.

TEST\SAMPLE			3050–2750 cm^{-1}	1525–120 cm^{-1}	1150–960 cm^{-1}	910–750 cm^{-1}
			\multicolumn{4}{c}{18 months}			
FDA	FANOVA	FP	≈0	≈0	≈0	≈0
		FB	≈0	≈0	≈0	≈0
VA	ANOVA		≈0	≈0	≈0	≈0
	Kruskal		≈0	≈0	≈0	≈0
			\multicolumn{4}{c}{12 months}			
FDA	FANOVA	FP	0.001	≈0	≈0	≈0
		FB	0.003	≈0	1.31×10^{-5}	≈0
VA	ANOVA		0.007	≈0	0.003	≈0
	Kruskal		0.008	2.23×10^{-6}	0.003	≈0
			\multicolumn{4}{c}{6 months}			
FDA	FANOVA	FP	≈0	≈0	≈0	≈0
		FB	≈0	≈0	≈0	1×10^{-4}
VA	ANOVA		≈0	≈0	≈0	≈0
	Kruskal		≈0	≈0	≈0	≈0

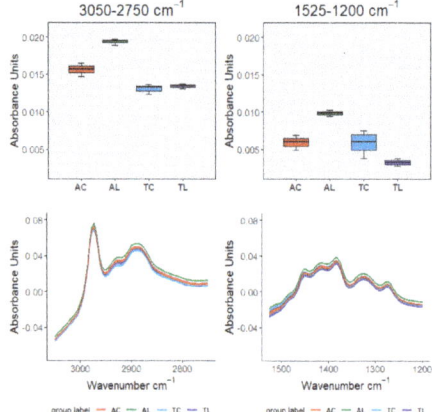

Figure 3. Plots of two of the four significant areas of the curves with an ageing time of 18 months. In the first row, vectorial analysis by means of boxplots. In the second row, functional data analysis (FDA) through curves in the studied interval. The wine spirit sample is divided into four groups (AC, AL, TC, and TL).

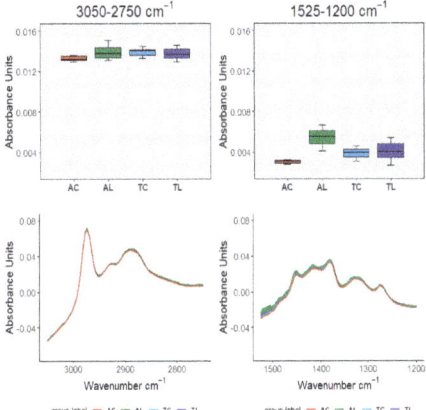

Figure 4. Plots of two of the four significant areas of the curves with an ageing time of 12 months. In the first row, vectorial analysis by means of boxplots. In the second row, FDA through curves in the studied interval. The wine spirit sample is divided into four groups (AC, AL, TC, and TL).

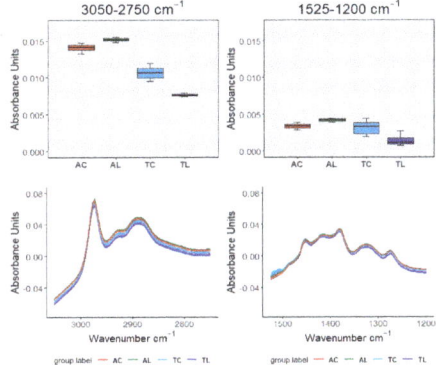

Figure 5. Plots of two of the four significant areas of the curves with an ageing time of 6 months. In the first row, vectorial analysis by means of boxplots. In the second row, FDA through curves in the studied interval. The wine spirit sample is divided into four groups (AC, AL, TC, and TL).

Secondly, the similarity between the three different ageing times was contrasted. This contrast is different depending on the ageing technology and kind of wood used (AC, AL, TC, and TL). Figure 6 shows the boxplots and curves of the sample resulting from alternative technology with chestnut wood. It can be seen that there is little difference between the three ageing times. Especially at the spectral region of 1150–960 cm^{-1} (second d column of Figure 6), in which the similarity hypothesis is rejected in the vector analysis. Instead, FDA is able to detect these differences between the curve samples (Table 5). This region is characteristic of the absorption bands assigned to C=O and C=C groups existing in furanic compounds, C–O stretch absorption bands related to ethanol and methanol, and C–C absorption bands also related to ethanol and some organic compounds such as sugars, phenols, alcohols, and esters previously reported, and all of them are important to differentiate the samples in this study. They are mainly identified at the peaks of the 1044 cm^{-1} and 1086 cm^{-1}, which are chiefly related to the presence of ethanol and methanol but also related to some organic compounds such as sugars, phenols, alcohols, and esters existing in minor concentration in the beverages. In addition, similarity in the other areas of the entire curve within the sample resulting from alternative technology with chestnut wood is rejected. In the case of the alternative technology, the wine spirits aged with

Limousin oak wood are very similar to those aged with chestnut wood but with more distance between the three ageing times (Figure 7), as observed in the chemical analysis. The similarity hypothesis is rejected in all areas and from both points of view (Table 5). Figure 8 shows the differentiation within the samples resulting from chestnut barrels. In this case, the two areas drawn from the whole curve are closer, but the differences between the three ageing times are greater. The hypothesis of similarity between the samples is rejected in all areas and from the vectorial and functional approach (Table 5). Finally, regarding the wine spirits aged in Limousin oak barrels, the 18 and 12-month samples show higher absorbance units than the 6-month sample (Figure 9). The spectrometric curves of the functional graph can be easily distinguished. There are significant reasons to reject the similarity between the three samples in all areas of the full curve from the two analyses (Table 5).

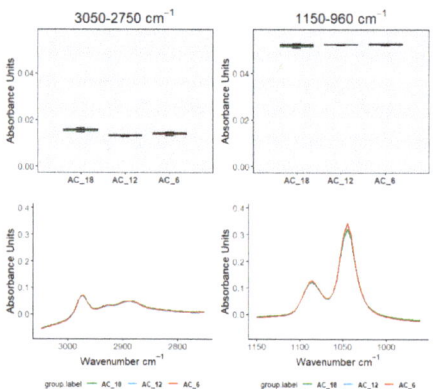

Figure 6. Plots of two of the four significant areas of the Alternative Chestnut (AC) curves. In the first row, vectorial analysis by means of boxplots. In the second row, FDA through curves in the studied interval. The wine spirit sample is divided into three groups depending on the ageing time (18, 12, and 6 months of ageing).

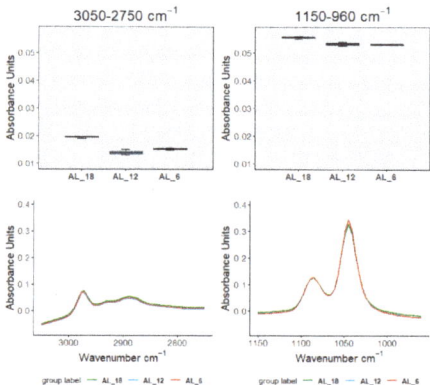

Figure 7. Plots of two of the four significant areas of the Alternative Oak (AL) curves. In the first row, vectorial analysis by means of boxplots. In the second row, FDA through curves in the studied interval. The wine spirit sample is divided into three groups depending on the ageing time (18, 12, and 6 months of ageing).

Table 5. Results of the similarity contrast between the three ageing times (18 months, 12 months, and 6 months), depending on the ageing technology. Functional results (FDA) are based on functional ANOVA (FANOVA) using two different tests (FP: permutation test based on a representation of the base function, FB: F test with a reduced bias estimation method). In addition, the results of the ANOVA and Kruskal's test representing the vectorial results (VA) are shown. All the results are *p*-values based on a 5% significance level.

TEST\SAMPLE			3050–2750 cm^{-1}	1525–120 cm^{-1}	1150–960 cm^{-1}	910–750 cm^{-1}
			Groups within AC			
FDA	FANOVA	FP	≈0	≈0	≈0	≈0
		FB	≈0	≈0	≈0	≈0
VA	ANOVA		≈0	≈0	0.032	≈0
	Kruskal		≈0	6.72×10^{-5}	0.214	1×10^{-4}
			Groups within AL			
FDA	FANOVA	FP	≈0	≈0	≈0	≈0
		FB	≈0	≈0	≈0	≈0
VA	ANOVA		≈0	≈0	≈0	≈0
	Kruskal		2.07e-06	4.14×10^{-6}	3.34×10^{-5}	3.71×10^{-6}
			Groups within TC			
FDA	FANOVA	FP	≈0	≈0	≈0	≈0
		FB	≈0	≈0	≈0	≈0
VA	ANOVA		≈0	≈0	≈0	≈0
	Kruskal		≈0	7.32×10^{-6}	≈0	≈0
			Groups within TL			
FDA	FANOVA	FP	≈0	≈0	≈0	≈0
		FB	≈0	≈0	≈0	≈0
VA	ANOVA		≈0	≈0	≈0	≈0
	Kruskal		1×10^{-4}	1.95×10^{-6}	≈0	≈0

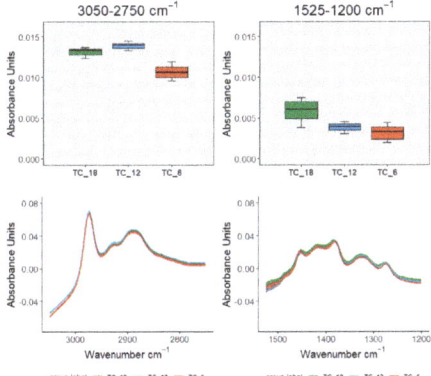

Figure 8. Plots of two of the four significant areas of the Traditional Chestnut (TC) curves. In the first row, vectorial analysis by means of boxplots. In the second row, FDA through curves in the studied interval. The wine spirit sample is divided into three groups depending on the ageing time (18, 12, and 6 months of ageing).

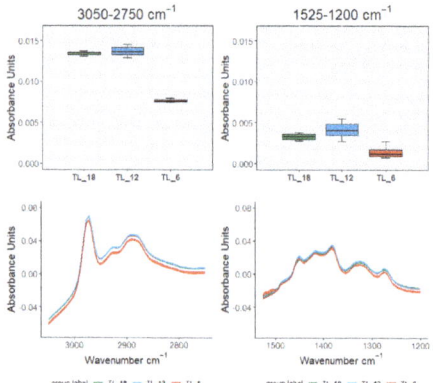

Figure 9. Plots of two of the four significant areas of the Traditional Oak (TL) curves. In the first row, vectorial analysis by means of boxplots. In the second row, FDA through curves in the studied interval. The wine spirit sample is divided into three groups depending on the ageing time (18, 12, and 6 months of ageing).

4. Discussion

The colour and total phenolic content acquired by the wine spirits during the ageing process (Table 1) are in line with the wine spirits' features observed in the first six months of ageing [8], and they are also in accordance with those of recent ageing studies of brandy [2,3]. The colour acquisition has been assigned to extraction, oxidation, and other physicochemical phenomena occurring during ageing [8,52]. Hence, more marked color development in the wine spirits aged by the alternative technology, and with chestnut wood, may have resulted from the direct application of oxygen [8] and/or from more oxygen transferred by this kind of wood due to its higher porosity [53], which may have favored such phenomena. Regarding the total phenolic content (TPI), a consistent behavior with the chromatic characteristics was found, reinforcing previous results [8]: there is a correlation between the enrichment in phenolic compounds and the colour acquired, which are induced by the ageing technology and the kind of wood.

Accordingly, the total content of low molecular weight compounds (phenolic acids, phenolic aldehydes, coumarins, and furanic aldehydes) concentrations determined by HPLC (Table 2) acted as distinctive feature of the aged wine spirits, which can be ordered as follows: AC > AL ~ TC > TL. Therefore, the pattern reported for the first six months of ageing [8] prevailed until the end of the ageing process (18 months).

Low molecular weight compounds, such as furfural, ellagic acid, vanillin, and coniferaldehyde, as aforementioned, stood out as markers of the ageing technology (Table 2). Similar results for these wood-derived phenolic compounds [1,54] were obtained in the beginning of ageing [8]. Their higher accumulation in wine spirits resulting from the alternative technology has been ascribed to a faster extraction from the wood, as described by other authors for the ageing of red wine with wood pieces [55,56]; the behavior of coniferaldehyde is explained by its sensitivity towards oxidation.

The evolution observed for the colour parameters over the ageing process was supported by the progressive increase of the total phenolic content (TPI) and particularly the low molecular weight compounds quantified by HPLC (sumHPLC), which reflected the release of the wood-derived compounds into the wine spirit together with a positive balance between their formation/degradation [4] in the liquid medium. The kinetics of individual compounds shows a continuous increment of ellagic acid content, regardless of the assay modality, which may have resulted from the release of this phenolic acid existing in the wood [1] and from the oxidation and hydrolysis of ellagitannins in the wine spirit during ageing [57,58]. The kinetics of vanillin presented a similar pattern, which was likely due to its release from the wood [59], lignin's hydroalcoholysis, and subsequent oxidation reactions that converted coniferaldehyde into vanillin in the wine spirit [4]. Interestingly, there was an increase of coniferaldehyde content between 6 and 12 months followed by a decrease until 18 months. This suggests that after 12 months of ageing, oxidation reactions prevailed over extraction and lignin's hydroalcoholys is causing an unbalance between coniferaldehyde formation and degradation. Similar results were previously noticed for wine spirits aged by the alternative technology but without micro-oxygenation and also by the traditional ageing in wooden barrels [4]. The evolution of furfural content was different from those of phenolic compounds. Indeed, its kinetics was closely related to the ageing technology: there was an increase from 6 to 12 months and then a decrease in wine spirits aged by the alternative technology; the opposite behavior occurred in wine spirits aged by the traditional one (TC and TL).

The statistical analyses performed during the ageing time reveals significant differences between 6 months and 12 months in a higher number of parameters than the observed between 12 months and 18 months. However, for the wine spirits aged by the alternative technology with Limousin staves significant differences between the three sampling times were observed. Regarding the alternative technology, these outcomes suggest that one year is enough/adequate to obtain a high-quality wine spirit aged with micro-oxygenation and chestnut staves as a consequence of higher pool of compounds and specific anatomical properties of this kind of wood compared with the Limousin oak wood [2,60].

The mathematical analysis performed with functional and vectorial approach corroborates the findings about colour parameters, total phenolic content, and low molecular weight compounds. According to the mathematical analysis, two main conclusions have been reached. First, there are significant reasons to reject the similarity between the wine spirits samples differentiated according to the ageing technology used. It was rejected in all areas of the full spectrometric curve for each of the three ageing times tested, both in vector and functional data analysis. Secondly, evidence was found to state that the samples of wine spirits with different ageing time are not similar. The similarity hypothesis between the three ageing times for each of the different technologies (traditional vs. alternative) used was rejected in all the areas of the spectrometric curve analysed.

On the other hand, FDA shows higher consistency than the vector analysis as observed by Martínez et al. [15,20]. In addition, to provide more information and accuracy, FDA can detect significant differences between groups that vector analysis cannot. This can be seen in the sample obtained by alternative technology with chestnut wood, specifically in the spectral region of 1150–960 cm^{-1} of the entire curve. It is attributed to the functional groups present in furanic compounds and also related to ethanol, methanol, and sugars, phenols, and esters existing in the wine spirit [1,4,5,8,14,49,51]. Vector analysis found no evidence to reject the similarity between the samples based on ageing time, but FDA, taking into account all correlated observations measured in the specific area, did.

5. Conclusions

The aged wine spirit composition and quality depend on the raw material, but mainly on the ageing technology, the wood species used in the ageing process, and the ageing time. Monitoring this quality over time with a fast methodology is very important for the industry.

This study showed a remarkable congruence between the analytical determinations (colour, total phenolic index, and low molecular weight compounds) and the FTIR-ATR/mathematical approach in the differentiation of wine spirits aged by the alternative (micro-oxygenation combined with wood pieces in stainless steel tanks) and the traditional (wooden barrels) technologies, using two kinds of wood (chestnut and oak), over the ageing time (6, 12, and 18 months). Additionally, the analytical results suggest that one year is enough/adequate to achieve a high-quality wine spirit aged with micro-oxygenation and chestnut staves.

FTIR-ATR with appropriate chemometric techniques, specifically functional data analysis, and vector analysis proved to be a powerful tool for an easier and faster monitoring of the wine spirit's ageing process. FDA showed higher consistency than the vector analysis, providing more information and accuracy; it detected significant differences between groups that the vector analysis did not detect.

It was clearly demonstrated in the sample resulting from the alternative technology with chestnut wood, specifically in the area 1150–960 cm^{-1} of the entire curve. Vector analysis found no evidence to reject similarity between the time ageing samples, but FDA, taking into account all correlated observation measured in the specific area, found it.

It was also possible to identify the more accurate spectral region to perform a calibration model to be applied by the wine spirit industry.

Author Contributions: Conceptualisation, O.A. and S.C.; methodology, O.A., S.C., M.M.C., P.E.O.; formal analysis, S.C., O.A., I.C., S.I.P.; Statistical analysis: M.M.C., P.E.O., O.A.; investigation, S.C., O.A., I.C., S.I.P.; resources, S.C.; O.A.; writing, O.A., M.M.C., S.C.; supervision, O.A.; project administration, S.C.; funding acquisition, S.C. All authors have read and agreed to the published version of the manuscript.

Funding: This research was funded by the Project CENTRO-04-3928-FEDER 0000281/Line "Validation of a new ageing technology for wine spirit from Lourinhã". This work is also funded by National Funds through FCT—Foundation for Science and Technology under the Project UIDB/05183/2020 (MED—Mediterranean Institute for Agriculture, Environment and Development) and the Project UIDB/00239/2020 (Centro de Estudos Florestais).

Acknowledgments: The authors thank João Pedro Catela, Nádia Santos, Manuela Gomes, Eugénia Gomes and Inês Antunes from Adega Cooperativa da Lourinhã, and Tanoaria J.M. Gonçalves for the technical support.

Conflicts of Interest: The authors declare no conflict of interest.

References

1. Canas, S. Phenolic Composition and Related Properties of Aged Wine Spirits: Influence of Barrel Characteristics. A Review. *Beverages* **2017**, *3*, 55. [CrossRef]
2. García-Moreno, M.V.; Sánchez-Guillén, M.M.; Mier, M.R.D.; Delgado-González, M.J.; Rodríguez-Dodero, M.C.; García-Barroso, C.; Guillén-Sánchez, D.A. Use of Alternative Wood for the Ageing of Brandy de Jerez. *Foods* **2020**, *9*, 250. [CrossRef] [PubMed]
3. Schwarz, M.; Rodríguez-Dodero, C.M.; Jurado, S.M.; Puertas, B.G.; Barroso, C.; Guillén, A.D. Analytical Characterization and Sensory Analysis of Distillates of Different Varieties of Grapes Aged by an Accelerated Method. *Foods* **2020**, *9*, 277. [CrossRef] [PubMed]
4. Canas, S.; Caldeira, I.; Belchior, A.P. Extraction/oxidation kinetics of low molecular weight compounds in wine brandy resulting from different ageing technologies. *Food Chem.* **2013**, *138*, 2460–2467. [CrossRef]
5. Caldeira, I.; Santos, R.; Ricardo-da-Silva, J.M.; Anjos, O.; Mira, H.; Belchior, A.P.; Canas, S. Kinetics of odorant compounds in wine brandies aged in different systems. *Food Chem.* **2016**, *211*, 937–946. [CrossRef]
6. Rodríguez-Solana, R.; Rodríguez-Freigedo, S.; Salgado, J.M.; Domínguez, J.M.; Cortés-Diéguez, S. Optimisation of accelerated ageing of grape marc distillate on a micro-scale process using a Box–Benhken design: Influence of oak origin, fragment size and toast level on the composition of the final product. *Aust. J. Grape Wine Res.* **2017**, *23*, 5–14. [CrossRef]
7. Gómez-Plaza, E.; Bautista-Ortín, A.B. Chapter 10—Emerging Technologies for Aging Wines: Use of Chips and Micro-Oxygenation. In *Red Wine Technology*; Morata, A., Ed.; Academic Press: Cambridge, MA, USA, 2019; pp. 149–162. [CrossRef]
8. Canas, S.; Caldeira, I.; Anjos, O.; Belchior, A. Phenolic profile and colour acquired by the wine spirit in the beginning of ageing: Alternative technology using micro-oxygenation vs traditional technology. *LWT Food Sci. Technol.* **2019**, *111*, 260–269. [CrossRef]
9. Llario, R.; Iñón, F.A.; Garrigues, S.; de la Guardia, M. Determination of quality parameters of beers by the use of attenuated total reflectance-Fourier transform infrared spectroscopy. *Talanta* **2006**, *69*, 469–480. [CrossRef]
10. Yadav, P.K.; Sharma, R.M. Classification of illicit liquors based on their geographic origin using Attenuated total reflectance (ATR)—Fourier transform infrared (FT-IR) spectroscopy and chemometrics. *Forensic Sci. Int.* **2019**, *295*, e1–e5. [CrossRef]
11. Nagarajan, R.; Mehrotra, R.; Bajaj, M.M. Quantitative analysis of methanol, an adulterant in alcoholic beverages, using attenuated total reflectance spectroscopy. *J. Sci. Ind. Res.* **2006**, *65*, 416–419.
12. Teodora Emilia, C.; Carmen, S.; Florinela, F.; FloricuÅ£a, R.; Raluca Maria, P.O.P.; Mira, F. Rapid Quantitative Analysis of Ethanol and Prediction of Methanol Content in Traditional Fruit Brandies from Romania, using FTIR Spectroscopy and Chemometrics. *Not. Bot. Horti Agrobot. Cluj-Napoca* **2013**, *41*. [CrossRef]

13. Wiśniewska, P.; Boqué, R.; Borràs, E.; Busto, O.; Wardencki, W.; Namieśnik, J.; Dymerski, T. Authentication of whisky due to its botanical origin and way of production by instrumental analysis and multivariate classification methods. *Spectrochim. Acta. Part A Mol. Biomol. Spectrosc.* **2017**, *173*, 849–853. [CrossRef] [PubMed]
14. Anjos, O.; Santos, A.J.A.; Estevinho, L.M.; Caldeira, I. FTIR–ATR spectroscopy applied to quality control of grape-derived spirits. *Food Chem.* **2016**, *205*, 28–35. [CrossRef] [PubMed]
15. Martínez Comesaña, M.; Martínez Mariño, S.; Eguía Oller, P.; Granada Álvarez, E.; Erkoreka González, A. A Functional Data Analysis for Assessing the Impact of a Retrofitting in the Energy Performance of a Building. *Mathematics* **2020**, *8*, 547. [CrossRef]
16. Martínez, J.; Ordoñez, C.; Matias, J.M.; Taboada, J. Determining noise in an aggregates plant using functional statistics. *Hum. Ecol. Risk Assess.* **2011**, *17*, 521–533. [CrossRef]
17. Müller, H.-G.; Sen, R.; Stadtmüller, U. Functional data analysis for volatility. *J. Econom.* **2011**, *165*, 233–245. [CrossRef]
18. López, M.; Martínez, J.; Matías, J.M.; Taboada, J.; Vilán, J.A. Shape functional optimization with restrictions boosted with machine learning techniques. *J. Comput. Appl. Math.* **2010**, *234*, 2609–2615. [CrossRef]
19. Sancho, J.; Pastor, J.J.; Martínez, J.; García, M.A. Evaluation of Harmonic Variability in Electrical Power Systems through Statistical Control of Quality and Functional Data Analysis. *Procedia Eng.* **2013**, *63*, 295–302. [CrossRef]
20. Martínez, J.; Pastor, J.; Sancho, J.; McNabola, A.; Martínez, M.; Gallagher, J. A functional data analysis approach for the detection of air pollution episodes and outliers: A case study in Dublin, Ireland. *Mathematics* **2020**, *8*, 225. [CrossRef]
21. Di Battista, T.; Fortuna, F. Functional confidence bands for lichen biodiversity profiles: A case study in Tuscany region (central Italy). *Stat. Anal. Data Min. ASA Data Sci. J.* **2017**, *10*, 21–28. [CrossRef]
22. Ruiz-Bellido, M.A.; Romero-Gil, V.; García-García, P.; Rodríguez-Gómez, F.; Arroyo-López, F.N.; Garrido-Fernández, A. Data on the application of Functional Data Analysis in food fermentations. *Data Brief* **2016**, *9*, 401–412. [CrossRef] [PubMed]
23. Ordoñez, C.; Martínez, J.; Matías, J.M.; Reyes, A.N.; Rodríguez-Pérez, J.R. Functional statistical techniques applied to vine leaf water content determination. *Math. Comput. Model.* **2010**, *52*, 1116–1122. [CrossRef]
24. Crawley, M.J. *The R Book*; John Wiley & Sons: Hoboken, NJ, USA, 2013.
25. Montgomery, D.C. *Design and Analysis of Experiments*; John Wiley & Sons, Inc.: Hoboken, NJ, USA, 2013.
26. Ostertagová, E.; Ostertag, O.; Kovác, J. Methodology and application of the Kruskal-Wallis test. *Appl. Mech. Mater.* **2014**, *611*, 115–120. [CrossRef]
27. Cetó, X.; Gutiérrez, J.M.; Gutiérrez, M.; Céspedes, F.; Capdevila, J.; Mínguez, S.; Jiménez-Jorquera, C.; del Valle, M. Determination of total polyphenol index in wines employing a voltammetric electronic tongue. *Anal. Chim. Acta* **2012**, *732*, 172–179. [CrossRef]
28. Canas, S.; Belchior, A.; Spranger, M.; Sousa, R. High-performance liquid chromatography method for analysis of phenolic acids, phenolic aldehydes, and furanic derivatives in brandies. Development and validation. *J. Sep. Sci.* **2003**, *26*, 496–502. [CrossRef]
29. Martínez, J.; Saavedra, Á.; García, P.J.; Piñeiro, J.I.; Iglesias, C.; Taboada, J.; Sancho, J.; Pastor, J. Air quality parameters outliers detection using functional data analysis in the Langreo urban area (Northern Spain). *Appl. Math. Comput.* **2014**, *241*, 1–10. [CrossRef]
30. Wang, J.L.; Chiou, J.M.; Müller, H.G. Functional Data Analysis. *Annu. Rev. Stat. Appl.* **2016**, *3*, 257–295. [CrossRef]
31. Kramosil, I.; Michálek, J. Fuzzy metrics and statistical metric spaces. *Kybernetika* **1975**, *11*, 336–344.
32. Ramsay, J.O.; Silverman, B.W. *Functional Data Analysis*, 2nd ed.; Springer: New York, NY, USA, 2005.
33. Cormier, E.; Genest, C.; Nešlehová, J.G. Using B-splines for nonparametric inference on bivariate extreme-value copulas. *Extremes* **2014**, *17*, 633–659. [CrossRef]
34. Kwok, W.Y.; Moser, R.D.; Jiménez, J. A Critical Evaluation of the Resolution Properties of B-Spline and Compact Finite Difference Methods. *J. Comput. Phys.* **2001**, *174*, 510–551. [CrossRef]
35. Piñeiro, J.I.; Torres, J.M.; García, P.J.; Alonso, J.R.; Muñiz, C.D.; Taboada, J. Analysis and detection of functional outliers in waterquality parameters from different automated monitoring stationsin the Nalón River Basin (Northern Spain). *Environ. Sci. Pollut. Res. Int.* **2015**, *22*, 387–396. [CrossRef] [PubMed]

36. Febrero, M.; Galeano, P.; Wenceslao, G.M. Outlier detection in functional data by depth measures, with application to identify abnormal NOx levels. *Environmetrics* **2008**, *19*, 331–345. [CrossRef]
37. Fraiman, R.; Muniz, G. Trimmed means for functional data. *TEST* **2001**, *10*, 419–440. [CrossRef]
38. Cuevas, A.; Febrero, M.; Fraiman, R. Robust estimation and classification for functional data via projection-based notions. *Comput. Stat.* **2007**, *22*, 481–496. [CrossRef]
39. Cuevas, A.; Febrero, M.; Fraiman, R. On the use of bootstrap for estimating functions with functional data. *Comput. Stat. Data Anal.* **2006**, *51*, 1063–1074. [CrossRef]
40. Millán-Roures, L.; Epifanio, I.; Martínez, V. Detection of Anomalies in Water Networks by Functional Data Analysis. *Math. Probl. Eng.* **2018**, *2018*, 13. [CrossRef]
41. Maturo, F.; Battista, T.; Fortuna, F. Parametric functional analysis of variance for fish biodiversity assessment. *J. Environ. Inform.* **2016**, *28*, 101–109. [CrossRef]
42. Cuesta-Albertos, J.A.; Febrero, M. A simple multiway ANOVA for functional data. *TEST* **2010**, *19*, 537–557. [CrossRef]
43. Zhang, J.-T. Analysis of Variance for Functional Data. In *A Chapman & Hall Book*; Press, C.R.C., Ed.; Taylor & Francis Group: Abingdon, UK, 2013; p. 412.
44. Aguilera, A.; Fortuna, F.; Escabias, M.; Battista, T. Assessing Social Interest in Burnout Using Google Trends Data. *Soc. Indic. Res.* **2019**, 1–13. [CrossRef]
45. Górecki, T.; Smaga, Ł. fdANOVA: An R software package for analysis of variance for univariate and multivariate functional data. *Comput. Stat.* **2019**, *34*, 571–597. [CrossRef]
46. Zhang, J.-T. Statistical inferences for linear models with functional responses. *Stat. Sin.* **2011**, *21*, 1431–1451. [CrossRef]
47. Górecki, T.; Smaga, Ł. A comparison of tests for the one-way ANOVA problem for functional data. *Comput. Stat.* **2015**, *30*, 987–1010. [CrossRef]
48. Moreira, J.; Santos, L. Spectroscopic interferences in Fourier transform infrared wine analysis. *Anal. Chim. Acta* **2003**, *513*, 263–268. [CrossRef]
49. Shurvell, H.F. Spectra—Structure Correlations in the Mid- and Far-Infrared. In *Handbook of Vibrational Spectroscopy*; John Wiley & Sons: Hoboken, NJ, USA, 2001.
50. Tarantilis, P.A.; Troianou, V.E.; Pappas, C.S.; Kotseridis, Y.S.; Polissiou, M.G. Differentiation of Greek red wines on the basis of grape variety using attenuated total reflectance Fourier transform infrared spectroscopy. *Food Chem.* **2008**, *111*, 192–196. [CrossRef]
51. Öztürk, B.; Yucesoy, D.; Ozen, B. Application of Mid-infrared Spectroscopy for the Measurement of Several Quality Parameters of Alcoholic Beverages, Wineand Raki. *Food Anal. Methods* **2012**, *5*, 1435–1442. [CrossRef]
52. Fujieda, M.; Tanaka, T.; Suwa, Y.; Koshimizu, S.; Kouno, I. Isolation and Structure of Whiskey Polyphenols Produced by Oxidation of Oak Wood Ellagitannins. *J. Agric. Food. Chem.* **2008**, *56*, 7305–7310. [CrossRef]
53. Carvalho, A. Identificação anatómica e caracterização física e mecânica das madeiras utilizadas no fabrico de quartolas para produção de aguardentes velhas de qualidade—Denominação Lourinhã. *Ciência Técnica Vitivinícola* **1998**, *13*, 71–105.
54. Martínez-Gil, A.; del Alamo-Sanza, M.; Sánchez-Gómez, R.; Nevares, I. Alternative Woods in Enology: Characterization of Tannin and Low Molecular Weight Phenol Compounds with Respect to Traditional Oak Woods. A Review. *Molecules* **2020**, *25*, 1474. [CrossRef]
55. Oberholster, A.; Elmendorf, B.L.; Lerno, L.A.; King, E.S.; Heymann, H.; Brenneman, C.E.; Boulton, R.B. Barrel maturation, oak alternatives and micro-oxygenation: Influence on red wine aging and quality. *Food Chem.* **2015**, *173*, 1250–1258. [CrossRef]
56. Laqui-Estaña, J.; López-Solís, R.; Peña-Neira, Á.; Medel-Marabolí, M.; Obreque-Slier, E. Wines in contact with oak wood: The impact of the variety (Carménère and Cabernet Sauvignon), format (barrels, chips and staves), and aging time on the phenolic composition. *J. Sci. Food Agric.* **2019**, *99*, 436–448. [CrossRef]
57. Karvela, E.; Makris, D.P.; Kefalas, P.; Moutounet, M. Extraction of phenolics in liquid model matrices containing oak chips: Kinetics, liquid chromatography-mass spectroscopy characterisation and association with in vitro antiradical activity. *Food Chem.* **2008**, *110*, 263–272. [CrossRef] [PubMed]
58. García-Estévez, I.; Alcalde-Eon, C.; Martínez-Gil, A.M.; Rivas-Gonzalo, J.C.; Escribano-Bailón, M.T.; Nevares, I.; del Alamo-Sanza, M. An Approach to the Study of the Interactions between Ellagitannins and Oxygen during Oak Wood Aging. *J. Agric. Food. Chem.* **2017**, *65*, 6369–6378. [CrossRef] [PubMed]

59. Martínez-Gil, A.; Cadahía, E.; Fernández de Simón, B.; Gutiérrez-Gamboa, G.; Nevares, I.; del Álamo-Sanza, M. Phenolic and volatile compounds in Quercus humboldtii Bonpl. wood: Effect of toasting with respect to oaks traditionally used in cooperage. *J. Sci. Food Agric.* **2019**, *99*, 315–324. [CrossRef] [PubMed]
60. Anjos, O.; Carmona, C.; Caldeira, I.; Canas, S. Variation of Extractable Compounds and Lignin Contents in Wood Fragments Used in the Aging of Wine Brandies. *BioResources* **2013**, *8*, 4484–4496. [CrossRef]

© 2020 by the authors. Licensee MDPI, Basel, Switzerland. This article is an open access article distributed under the terms and conditions of the Creative Commons Attribution (CC BY) license (http://creativecommons.org/licenses/by/4.0/).

Article

Outliers Detection Models in Shewhart Control Charts; An Application in Photolithography: A Semiconductor Manufacturing Industry

Ishaq Adeyanju Raji [1], Muhammad Hisyam Lee [1,*], Muhammad Riaz [2], Mu'azu Ramat Abujiya [3] and Nasir Abbas [2]

1. Department of Mathematical Sciences, Universiti Teknologi Malaysia, Skudai 81310, Malaysia; arishaq2@graduate.utm.my
2. Department of Mathematics and Statistics, King Fahd University of Petroleum and Minerals, Dhahran 31261, Saudi Arabia; riazm@kfupm.edu.sa (M.R.); nasirabbas@kfupm.edu.sa (N.A.)
3. Preparatory Year Mathematics Program, King Fahd University of Petroleum and Minerals, Dhahran 31261, Saudi Arabia; abujiya@kfupm.edu.sa
* Correspondence: mhl@utm.my; Tel.: +60-7-553-0264

Received: 10 April 2020; Accepted: 21 May 2020; Published: 25 May 2020

Abstract: Shewhart control charts with estimated control limits are widely used in practice. However, the estimated control limits are often affected by phase-I estimation errors. These estimation errors arise due to variation in the practitioner's choice of sample size as well as the presence of outlying errors in phase-I. The unnecessary variation, due to outlying errors, disturbs the control limits implying a less efficient control chart in phase-II. In this study, we propose models based on Tukey and median absolute deviation outlier detectors for detecting the errors in phase-I. These two outlier detection models are as efficient and robust as they are distribution free. Using the Monte-Carlo simulation method, we study the estimation effect via the proposed outlier detection models on the Shewhart chart in the normal as well as non-normal environments. The performance evaluation is done through studying the run length properties namely average run length and standard deviation run length. The findings of the study show that the proposed design structures are more stable in the presence of outlier detectors and require less phase-I observation to stabilize the run-length properties. Finally, we implement the findings of the current study in the semiconductor manufacturing industry, where a real dataset is extracted from a photolithography process.

Keywords: average run length; control chart; median absolute deviation; outlier; photolithography; Shewhart; Tukey

1. Introduction

The two salient tools of statistical process control (SPC) are memory and memory-less control charts. The memory-less control charts are most suitable for large shift, while the memory-control charts are used to monitor moderate and small shifts. The prominent form of memory-less control chart for location monitoring is the Shewhart $\overline{X}$ control chart. In general, control charts-irrespective of the magnitude they measure-operate in two phases: phase-I, the prospective stage from which the control limits are obtained; phase-II, where we monitor the process and correct the unnatural causes of variation whenever they occur (cf. [1]). In phase-I we estimate the control limits using the parameters of the process under study which, in reality, are seldom known. The amount of data employed in phase-I for estimating process parameters varies from one practitioner to the other. As a result, this variability affects the chart performance in the monitoring stage i.e., phase-II. (see for example [2–6]).

Furthermore, the amount of data employed in estimating the process parameters does affect the accuracy of the chart, as well as its limits. As we all know, the larger the sample size, the closer we are to the parameter. Therefore, increasing the sample size used for estimating the parameters should be the remedy to this shortcoming, but there is a limit to which we can increase sample sizes in real-life situations. As a result, the Shewhart chart, like any other chart, loses its performance and credibility. The depth of the loss depends on the efficacy of the parameter estimation and sample size employed in phase-I.

The presence of outlying/extreme values in the phase-I dataset can affect the performance of the control chart. The insufficiency of the phase-I estimates could be a result of extreme sample points in the sample, and not necessarily the size of the sample (see [5,7]). The easiest remedy for the extreme values is to drop such a sample and pick another one, but this is not appropriate for small sample data. Therefore, there is a need to screen the extreme values to improve the overall performance of the control chart.

Over the years, researchers have studied different types of robust outlier detection models in a series of control charts to enhance their performance. Examples include [8–12]. These outlier detectors require the data to be from normal distribution such as the Student-type and Grubbs-type detectors. However, for a non-normal dataset, the Tukey's and median absolute deviation (MAD) outlier detection models are more accurate and robust since they are independent of mean and standard deviation. (see [13–19]). SPC is widely applied and implemented in various sectors; health, industrial, manufacturing and every service-rendering sector. Control charts, however, are most applied in manufacturing industry, with semiconductors as a case study. Semiconductor manufacturing processes are prone to high chances of assignable cause of variations, due to machine breakdown, multiple products, re-entrant flows, batching processes etc. [20]. Researchers have employed SPC in solving these recurring challenges in this industry (see [21–24]). The proposed charts in this study are applied in photolithography, a semiconductor manufacturing process.

In this article, we study the effects of parameter estimation on the Shewhart $\overline{X}$ chart for normal and non-normal environments. We also study the effect of outliers on the reliability of the control charts and the process parameters are estimated. Furthermore, we propose non-parametric outlier detectors, namely: the robust Tukey and MAD outlier detection models in designing the basic control chart structure. A fair comparison between the two-outlier detection models is also made. We achieve all of these using average run length (ARL) and standard deviation run length (SDRL) as the performance measures.

The remainder of this article is as follows: the next Section entails the methodologies employed for the study; briefing the overview of the Shewhart $\overline{X}$ control chart when the parameters are known and unknown, alongside the performance measure properties adopted in this study; the variability in Shewhart chart performance due to phase-I estimation; a scenario for the presence of outliers in the design structure of Shewhart chart, and its effect; incorporating the Turkey and MAD outlier detection models in the design structure of the Shewhart chart as remedies for rectifying the presence of outliers; Section 3 gives a concise and precise description of the simulation results. In Section 4, a detailed comparison of the results is presented; while Section 5 provides an illustrative example with a real life dataset; finally a concluding remark and future recommendations are given in Section 6.

2. Methodology

In this section, we give details of the Shewhart control chart for normal and non-normal environments. The known and unknown parameter scenarios, the practitioner–practitioner variation in the estimation stage, the presence of outliers/extreme values in the estimation sample, and incorporating some outlier detection models in the Shewhart chart are all discussed in the following subsections.

2.1. Overview of the Shewhart Control Chart

Let Y_{ij} $i = 1, 2, \ldots, n$ and $j = 1, 2, \ldots$ represent a ith observation from jth sample of an ongoing (continuous) process. Further Y_{ij} follows a normal distribution with mean $\mu_0 + \delta\sigma_0$ and variance σ_0^2 i.e., $Y_{ij} \sim N(\mu_0 + \delta\sigma_0, \sigma_0^2)$. The process is said to be in the in-control (IC) state if $\delta = 0$, and out-of-control (OoC) otherwise. A default Shewhart set-up monitors a process by plotting the sample mean ($\overline{Y}_i = 1/n \sum_{j=1}^{n} Y_{ij}$) of Y_{ij} against the following control chart limits.

$$\text{UCL} = \mu_0 + L\frac{\sigma_0}{\sqrt{n}}, \quad \text{LCL} = \mu_0 - L\frac{\sigma_0}{\sqrt{n}} \quad (1)$$

where UCL and LCL denote the upper and lower control limits, respectively. Limits in (1) are useful when the parameters (μ_0 and σ_0^2) of the process are known. However, when they are unknown, their respective unbiased estimators from the phase-I are used, and the resulting control chart structures will be in estimated form.

For phase-I, let Y_{il} represents ith observation from lth random sample $\forall\ i = 1, 2, 3, \ldots, n$ and $l = 1, 2, 3, \ldots, m$, regarded to be under statistically IC state. It is good to mention here that the choice of m and n varies from one practitioner-to another. Therefore, it affects the accuracy of the control limits implying an influenced ARL in phase-II. The unbiased estimators for the parameters μ and σ of an IC process are defined as:

$$\hat{\mu}_0 = (1/m) \sum_{l=1}^{m} \overline{Y}_l$$
$$\hat{\sigma}_0 = (1/mC_4) \sum_{l=1}^{m} S_l \quad (2)$$

where $\overline{Y}_l = \frac{\sum_{i=1}^{n} Y_{il}}{n}$, $S_l = \sqrt{\frac{\sum_{i=1}^{n}(Y_{il}-\overline{Y}_l)^2}{n-1}}$ and $c_4 = \frac{\sqrt{2/(n-1)}\Gamma(n/2)}{\Gamma[(n-1)/2]}$ is the bias correction constant. Subsequently, the resulting control limits in (1) are modified to the following:

$$\widehat{\text{UCL}} = \frac{\sum_{l=1}^{m}\overline{Y}_l}{m} + \hat{L}\frac{\sum_{l=1}^{m} S_l}{mC_4\sqrt{n}}, \quad \widehat{\text{LCL}} = \frac{\sum_{l=1}^{m}\overline{Y}_l}{m} - \hat{L}\frac{\sum_{l=1}^{m} S_l}{mC_4\sqrt{n}} \quad (3)$$

In phase-II, $\overline{Y}_l$s are plotted against the control limits in (3) and the chart is said to have given an OoC signal if any value of $\overline{Y}_l$ is plotted outside the limits. Here, the sample number at which the statistic is plotted outside the limits is recorded as run length (RL). RL is an important variable in measuring the performance of control charts in general, and the Shewhart is not an exception. The most widely used property of RL is ARL, which is the average number of samples observed before the chart sends an OoC signal. Mathematically, $\text{ARL} = \sum_{k=1}^{s} \text{RL}_k/s$ where s is the number of RLs recorded. In addition to ARL, standard deviation of the RL (SDRL) gives more information about the behavior of the RL variable in evaluating the performance of a control chart. Furthermore, the ARL is of two types i.e., the IC ARL, denoted as ARL_0 and the OoC ARL, referred to as ARL_1. ARL_0 is expected to be sufficiently large enough to avoid false alarms. On the other hand, ARL_1 is anticipated to be sufficiently small to enable the process to send a signal as soon as there is a shift in the process parameter(s).

2.2. Variability in the Shewhart Chart Performance

In this section, we explain the effect of the practitioner to practitioner variability on the Shewhart chart, both in normal and non-normal distribution, by using the Monte Carlo simulation approach. See ([25–29]) for more information about the effect of sample size and practitioners' variability. To achieve this aim, we develop an algorithm in R programing language to simulate the Shewhart chart environment, using the standard Shewhart chart as our benchmark and reference point. The $\overline{X}$ chart has a control limits width determinant L that influences RL properties. We use the standard $L = 3$, that corresponds to the $\text{ARL}_0 = 370$ (see [1] for more details). Without any loss of generality, we generate random samples from a standard normal distribution $N(\mu = 0, \sigma = 1)$, each of sample size $n = 5$, assuming the process parameters are known. While for the non-normal distribution, we considered the t-distribution with degrees of freedom $v = 5, 25$, and 100. Since all the three categories of v exhibit the same pattern, we report only the results for $v = 100$. In both environments, normal and t-distributions, we set up the chart limits as given in Equation (1) and plot the sample means against the UCL and LCL. As soon as a value of $\overline{Y}_j$ is plotted outside the limits, RL is recorded and saved. The process is iterated 10^5 times to get ARL and SDRL.

For the unknown parameters, we estimate the parameter from phase-I. The number of samples employed for the estimation differs from on practitioner to another and so does the accuracy of the charts in phase-II. To depict that, we estimated both μ_0 and σ_0 from different number of in-control phase-I samples i.e., $m = 25, 50, 100, 250, 500$ and 1000 each of sample size $n = 5$. The estimated parameters $\hat{\mu}_0$ and $\hat{\sigma}_0$ from the phase-I IC stage are, therefore, used in the same algorithm instead of μ_0 and σ_0 respectively. Subsequently the parameter L, changes as the amount of phase-I samples changes. The corresponding L's for the different m's are $L = 2.962, 2.983, 2.9925, 2.997, 2.999$, and 3 respectively for the normal distribution, and $L = 2.974, 2.995, 3.005, 3.010, 3.012$, and 3.012 respectively for the t-distribution of $v = 100$. These L's are determined through simulations to obtain $ARL_0 = 370$. We carry out the simulation with different level of shifts δ ranging from 0 to 5 i.e., $\delta \in (0, 0.5, 5)$, as shown in Tables 1 and 2.

Table 1. Average run length (ARL) of the Shewhart chart with estimated parameters for standard normal and t (v = 100) distributions.

ARL	Standard Normal Distribution						T-Distribution v = 100					
δ/m	25	50	100	250	500	1000	25	50	100	250	500	1000
0	370.93	369.66	370.43	369.09	371.10	370.98	370.67	370.00	369.58	370.89	369.63	370.00
0.5	190.76	173.92	165.35	159.25	157.00	156.54	194.06	177.76	167.85	161.75	160.48	159.04
1	53.91	48.74	46.37	44.73	44.41	44.21	55.14	50.13	47.75	46.26	46.02	45.32
1.5	17.19	16.05	15.52	15.20	15.09	15.04	17.70	16.53	16.06	15.65	15.65	15.52
2	6.87	6.59	6.46	6.35	6.33	6.33	7.05	6.80	6.65	6.57	6.54	6.57
2.5	3.40	3.33	3.29	3.26	3.25	3.25	3.52	3.44	3.40	3.36	3.35	3.35
3	2.05	2.03	2.02	2.01	2.00	2.00	2.10	2.08	2.06	2.06	2.05	2.05
3.5	1.47	1.46	1.45	1.45	1.45	1.45	1.50	1.49	1.48	1.48	1.48	1.47
4	1.20	1.19	1.19	1.19	1.19	1.19	1.21	1.21	1.21	1.21	1.20	1.20
4.5	1.08	1.08	1.07	1.07	1.07	1.07	1.09	1.08	1.08	1.08	1.08	1.08
5	1.03	1.02	1.02	1.02	1.02	1.02	1.03	1.03	1.03	1.03	1.03	1.03
L	2.962	2.983	2.9925	2.997	2.999	3	2.974	2.995	3.005	3.010	3.012	3.012

Table 2. Standard deviation of the run length (SDRL) of the Shewhart chart with estimated parameters for standard normal and t (v = 100) distributions.

SDRL	Standard Normal Distribution						T-Distribution v = 100					
δ/m	25	50	100	250	500	1000	25	50	100	250	500	1000
0	601.54	473.20	420.75	387.76	379.99	375.44	601.69	468.55	416.69	389.69	376.91	373.30
0.5	333.88	232.92	193.19	169.15	161.43	158.10	340.19	241.98	197.21	172.49	164.11	160.46
1	88.35	61.71	52.00	46.39	45.00	44.24	91.10	64.04	53.93	47.96	46.61	45.29
1.5	24.05	18.52	16.40	15.20	14.85	14.67	25.00	19.21	16.92	15.71	15.33	15.20
2	8.24	6.88	6.32	5.97	5.90	5.85	8.73	7.10	6.48	6.22	6.11	6.11
2.5	3.40	3.04	2.86	2.76	2.72	2.71	3.60	3.16	2.99	2.86	2.84	2.83
3	1.66	1.54	1.48	1.44	1.43	1.42	1.70	1.59	1.54	1.49	1.48	1.47
3.5	0.90	0.85	0.83	0.81	0.81	0.81	0.94	0.89	0.86	0.84	0.84	0.84
4	0.51	0.49	0.48	0.48	0.48	0.48	0.54	0.51	0.51	0.50	0.50	0.50
4.5	0.30	0.29	0.28	0.28	0.28	0.28	0.32	0.31	0.30	0.29	0.29	0.29
5	0.17	0.16	0.16	0.16	0.16	0.16	0.18	0.17	0.17	0.17	0.17	0.17
L	2.962	2.983	2.9925	2.997	2.999	3	2.974	2.995	3.005	3.010	3.012	3.012

2.3. Presence of Outliers in the Shewhart Chart with Estimated Parameters

Although the estimation of the unknown parameters in phase-I samples plays its role on the efficiency of the control chart in phase-II. The drop in the efficacy of the chart performance is not limited to this fact alone, rather it extends to presence of outlying/extreme values in the phase-I samples.

In this Section, we study the effect of outliers in the phase-I samples on the performance and accuracy of the Shewhart chart. Here, through Monte Carlo simulation, we generate the m phase-I samples from a mixture distribution i.e., $(1-\alpha)100\%$ from assumed (normal or t-distribution) and the remaining $\alpha 100\%$ from a chi-square distribution with n degrees of freedom denoted by $\chi^2_{(n)}$. Subsequently, the estimated parameters emerging from the m samples have an extreme values effect on the control chart in phase-II. That is, each observation of the phase-I sample is generated from the following expression:

$$(1-\alpha)N(\mu,\sigma^2) + \alpha\left[N(\mu,\sigma^2) + w\,\chi^2_{(n)}\right] \text{ or }$$
$$(1-\alpha)t(v) + \alpha\left[t(v) + w\,\chi^2_{(n)}\right] \quad (4)$$

where $\alpha > 0$, is the probability of having a multiple of $\chi^2_{(n)}$ added to the assumed distribution, serving as the outliers in the samples. In addition, $w \geq 1$ is the magnitude of the outlier. We develop an algorithm from the R language, similar to that in Section 2.2, but the samples are from the environment described in (4). We set $\mu = 0$, $\sigma^2 = 1$, $v = 100$, $w = 3$, and $\alpha \in [0, 0.01]$. We design the Shewhart chart using the same parameters L and m as in Section 2.2.

In general, the pattern exhibited by the RL properties implies the following:

- Increasing the m phase-I samples in the presence of outliers, gets the ARL_0's closer to the theoretical values.
- Reducing the value of α, the percentage of outliers present in the m samples also brings the ARL_0's closer to the theoretical values.

Unfortunately, neither of the two suggested remedies is practicable in real life. Thus, we propose outliers detecting structures through the robust Turkey and MAD detection models.

2.4. Shewhart Chart with Outlier Detection Models

In the section, we propose two outlier-detecting models as remedy to the issues raised in Sections 2.2 and 2.3. The Tukey and the MAD model-based Shewhart charts. Their procedures applied in parallel to the Shewhart chart are described in the sub sections below:

2.4.1. The Tukey Shewhart Control Chart

For the phase-I samples, $\widetilde{Y}$ be the median of all $m \times n$ observations. For any observation y_o if $\left|y_o - \widetilde{Y}\right| > p \times IQR$, then y_o is declared an outlier. Here $IQR = Q_3 - Q_1$ is the inter-quartile range of the sample. Q_3 and Q_1 are the third and first quartiles, respectively, of all $m \times n$ phase-I observations. The constant p on the other hand is the confidence factor of the Tukey's detector, commonly chosen between 1.5 and 3.0. The confidence factor should be carefully chosen, and not too small, to avoid over detection. Also it should not be too large, to prevent under detection [18]. In this study, we choose $p = 2.2$. Applying the same algorithm, parameters and limits employed in Section 2.2, we incorporate the Tukey outlier-detector model on the phase-I samples to screen out the extreme values present there in. Then we compute the IC ARL and SDRL values for the Shewhart chart based on the Tukey model in phase-II, when the parameters are estimated.

2.4.2. The Median Absolute Deviation (MAD) Shewhart Control Chart

We define median absolute deviation (MAD) as the deviation of the dataset about the median as $\text{MAD} = median\left(\left|Y_{il} - \widetilde{Y}\right|\right)/0.6574$. Then it follows, that any observation y_o from the sample that falls outside the expression $\widetilde{Y} \pm b * \text{MAD}$, is declared an outlier. Here b is the outlier detecting constant and chosen 3.642 so that the percentage of screening by MAD is the same as Tukey. This has been done to keep the comparison between two outlier detectors valid [19].

Furthermore, it is worth distinguishing between outlying and OoC sample points. The former emerges from $m$phase-I samples, which are used to construct the control limits for the monitoring stage; phase-II; while the latter are the sample points that fall beyond the control limits in phase-II. Therefore, the presence of outlying sample points in phase-I leads to wider control limits, rendering the control charts less effective. A flowchart summarizing the procedure is depicted in Figure 1.

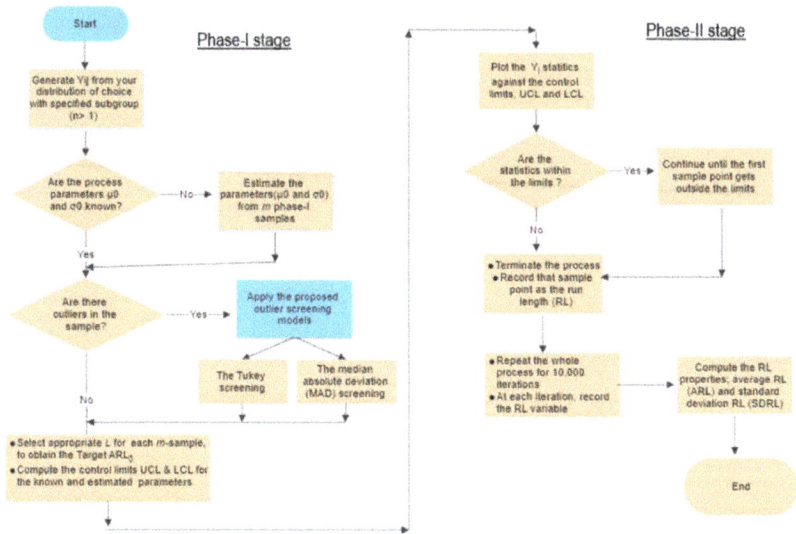

Figure 1. Flowchart of the procedures of proposed Shewhart control chart.

3. Results

In this section, we provide the results of the methodologies discussed in Section 2. These results are presented in three folds, so is the discussion in the next section.

3.1. Practitioners' Estimation Variability

Here, through the simulation results of the algorithm explained in Section 2.2, we observe the variability that appears in the Shewhart control chart due to different choices of sample size m, amongst practitioners. Tables 1 and 2 depict the Shewhart chart whose parameters, both mean and variance, are estimated from m phase-I samples for both normal and non-normal distributions. It is evident from the result, the effect of parameter estimation on the performance of the chart. The ARL_0s when $\delta = 0$, are clustering around the target 370 with their respective L's. However, when $\delta \neq 0$, we observe that the smaller m becomes, the less effective the Shewhart chart performance. The ARL_1's are expected to be sufficiently small in order to detect any drift in the ongoing process, but as m gets smaller, ARL_1's get bigger. Which implies the chart is less sensitive in identifying the presence of shifts in the ongoing process early enough. Another noticeable effect of the parameter estimation on the Shewhart chart is the decrement in the limits L, as m reduces. This should be recorded as an edge if the corresponding phase-II charts detects shift earlier than when the parameters are known.

3.2. Effect of Outliers on the Shewhart Control Charts

In Tables 3 and 4, we present the simulation results of environment (4) discussed in Section 2.3. From these results, the gross impact of outliers in the phase-I samples on the performance of the Shewhart chart cannot be over emphasized. Having seen the pattern of the IC and OoC RL properties in Tables 1 and 2, in order to save space, we restrict the performance evaluation to the IC RL properties. That is, considering the case when $\delta = 0$ only. From Tables 3 and 4, when $\alpha = 0$, in the absence of outlier, the ARL_0's are clustering around its target 370, irrespective of the amount of phase-I sample m. However, when $\alpha > 0$, the ARL_0's deviate from the target, vigorously. As the amount of phase-I samples m reduces, and the percentage of outliers present in the samples α increases, the more the ARL_0's deviate from the target. Similarly the pattern of the SDRL, even more.

Table 3. ARL of the Shewhart chart in the presence of outliers with estimated parameters for standard normal and t ($v = 100$) distributions.

ARL	Standard Normal Distribution						T-Distribution $v = 100$					
α/m	25	50	100	250	500	1000	25	50	100	250	500	1000
0	371.20	369.51	370.29	369.59	370.72	370.85	370.65	369.57	369.28	370.85	369.11	370.87
0.001	424.76	401.49	392.98	389.92	388.74	389.13	418.21	394.60	386.19	385.76	385.04	381.04
0.002	484.38	435.14	417.35	410.86	409.61	408.88	478.03	424.90	408.83	405.04	403.56	398.94
0.003	540.09	469.66	440.42	431.91	429.47	428.70	534.47	458.93	435.77	424.67	424.75	422.44
0.004	602.97	507.51	469.33	452.76	450.30	449.72	587.12	500.66	457.98	443.64	444.15	437.50
0.005	664.58	545.85	494.67	477.03	471.98	471.09	650.79	528.62	489.31	469.04	461.71	461.22
0.006	731.55	588.72	525.81	500.69	495.46	493.34	718.10	569.85	512.77	491.40	486.42	481.51
0.007	788.48	629.88	554.41	525.13	519.74	516.88	758.54	604.64	539.10	514.95	509.78	504.97
0.008	861.51	673.43	586.11	551.83	544.06	541.56	837.89	652.55	578.41	541.64	531.29	528.75
0.009	931.14	726.12	618.88	579.38	569.55	565.85	916.99	693.94	606.71	568.85	560.07	552.75
0.01	996.35	773.78	654.18	607.70	596.56	592.55	953.61	745.34	634.00	593.84	586.09	575.08
L	2.962	2.983	2.9925	2.997	2.999	3	2.974	2.995	3.005	3.010	3.012	3.012

Table 4. SDRL of the Shewhart chart in the presence of outliers with estimated parameters for standard normal and t ($v = 100$) distributions.

SDRL	Standard Normal Distribution						T-Distribution $v = 100$					
α/m	25	50	100	250	500	1000	25	50	100	250	500	1000
0	607.93	473.75	419.35	388.27	379.67	375.44	595.89	468.00	418.16	382.43	375.42	369.72
0.001	1238.32	664.83	467.76	414.51	400.57	394.71	1150.79	635.34	463.74	411.63	395.24	385.68
0.002	1704.83	838.46	515.41	441.77	423.19	415.18	1802.06	760.16	508.22	433.04	416.23	406.03
0.003	2041.64	991.95	568.33	468.17	446.51	436.26	2096.30	913.63	558.95	460.42	441.81	431.04
0.004	2410.92	1176.26	630.64	497.30	469.60	459.66	2354.27	1184.84	606.26	486.65	462.63	447.94
0.005	2689.45	1294.24	687.44	528.00	494.90	481.60	2626.16	1192.76	683.26	517.24	485.89	473.05
0.006	2994.88	1502.07	756.39	559.42	521.35	505.08	2871.17	1449.52	699.03	546.74	512.57	494.30
0.007	3193.80	1637.48	822.18	592.02	549.29	530.76	2956.37	1514.95	757.01	573.97	540.59	519.76
0.008	3480.50	1799.40	888.29	627.68	578.43	557.35	3370.66	1664.42	912.65	618.19	564.09	544.14
0.009	3772.38	1994.86	959.76	664.49	607.34	582.18	3703.48	1766.87	955.78	655.56	594.72	571.46
0.01	4012.72	2139.41	1028.74	702.52	638.62	612.00	3823.24	2009.14	980.24	685.94	628.22	591.55
L	2.962	2.983	2.9925	2.997	2.999	3	2.974	2.995	3.005	3.010	3.012	3.012

3.3. Improvement of Tukey and MAD Outlier Detection Models on Shewhart Chart Performance

While incorporating the procedures in Sections 2.4.1 and 2.4.2, the simulation results are presented in Tables 5–8 respectively. Tables 5 and 7 represents the ARL result for Tukey and MAD outlier detection models respectively, as Tables 6 and 8 are the corresponding SDRL results. The effect of these detection models are noticed as ARLs and SDRLs are closer to when there is an absence of outliers or even better.

Table 5. ARL of the Shewhart chart with Tukey outlier detection for standard normal and t ($v = 100$) distributions.

ARL	Standard Normal Distribution						T-Distribution $v = 100$					
a/m	25	50	100	250	500	1000	25	50	100	250	500	1000
0	363.24	363.44	364.17	364.36	365.34	366.80	358.67	358.71	357.97	358.10	359.64	357.52
0.001	365.71	366.33	366.48	366.83	367.56	368.22	361.59	360.26	357.98	362.04	359.85	360.39
0.002	367.81	367.30	368.51	368.71	369.89	366.67	364.59	363.25	359.73	359.63	363.42	361.61
0.003	370.30	370.04	370.91	371.14	370.90	371.03	365.85	365.75	364.12	361.29	362.85	362.73
0.004	372.35	371.48	372.44	372.76	372.72	374.13	368.84	365.44	365.73	365.64	365.87	366.25
0.005	376.13	374.05	374.49	373.95	375.73	377.23	370.27	368.13	366.95	366.59	366.89	367.05
0.006	377.97	376.06	376.96	376.61	376.88	379.44	372.27	368.44	368.28	371.20	371.40	370.77
0.007	380.13	378.38	379.03	378.98	379.15	379.44	374.78	372.44	371.54	370.82	370.45	371.76
0.008	383.48	380.17	380.51	380.63	380.56	381.29	378.04	372.69	372.06	373.31	376.35	374.45
0.009	386.35	383.02	383.50	382.73	383.29	382.84	377.58	374.74	371.66	375.88	375.09	372.29
0.01	388.82	384.62	385.57	383.80	384.95	386.39	381.84	377.80	378.17	377.29	378.48	374.09
L	2.962	2.983	2.9925	2.997	2.999	3	2.974	2.995	3.005	3.010	3.012	3.012

Table 6. SDRL of the Shewhart chart with Tukey outlier detection for standard normal and t ($v = 100$) distributions.

SDRL	Standard Normal Distribution						T-Distribution $v = 100$					
a/m	25	50	100	250	500	1000	25	50	100	250	500	1000
0	596.32	468.02	414.04	383.16	374.68	368.66	607.33	467.10	411.17	377.81	369.81	360.34
0.001	600.30	472.43	417.55	385.96	377.25	372.35	612.79	467.20	406.53	380.03	366.57	366.34
0.002	607.28	476.91	419.54	388.33	379.39	371.64	627.84	471.38	410.10	376.44	372.50	365.67
0.003	615.99	477.95	422.41	389.95	380.10	375.45	617.88	474.03	415.42	378.88	374.34	366.23
0.004	615.39	481.58	423.16	392.96	382.24	377.77	636.02	476.02	416.09	385.14	377.04	371.24
0.005	625.95	486.00	427.29	393.54	385.71	380.37	650.31	480.25	420.69	387.36	377.24	370.45
0.006	636.60	488.82	430.22	396.23	386.61	383.77	637.39	483.89	423.95	390.68	383.17	376.49
0.007	646.05	490.89	433.11	399.86	389.30	384.98	630.84	488.73	429.10	390.52	381.42	377.88
0.008	656.27	496.12	435.44	400.84	391.44	385.54	654.64	489.53	430.19	391.87	386.60	378.02
0.009	654.98	500.43	439.24	403.24	393.19	387.01	675.27	499.78	423.47	395.23	385.19	380.11
0.01	669.83	502.93	441.89	404.61	395.16	390.51	672.71	493.77	436.38	399.23	390.17	380.26
L	2.962	2.983	2.9925	2.997	2.999	3	2.974	2.995	3.005	3.010	3.012	3.012

Table 7. ARL of the Shewhart chart with median absolute deviation (MAD) outlier detection for standard normal and t ($v = 100$) distributions.

ARL	Standard Normal Distribution						T-Distribution $v = 100$					
a/m	25	50	100	250	500	1000	25	50	100	250	500	1000
0	363.24	363.44	364.17	364.36	365.34	366.80	358.67	358.71	357.97	358.10	359.64	357.52
0.001	365.71	366.33	366.48	366.83	367.56	368.22	361.59	360.26	357.98	362.04	359.85	360.39
0.002	367.81	367.30	368.51	368.71	369.89	366.67	364.59	363.25	359.73	359.63	363.42	361.61
0.003	370.30	370.04	370.91	371.14	370.90	371.03	365.85	365.75	364.12	361.29	362.85	362.73
0.004	372.35	371.48	372.44	372.76	372.72	374.13	368.84	365.44	365.73	365.64	365.87	366.25
0.005	376.13	374.05	374.49	373.95	375.73	377.23	370.27	368.13	366.95	366.59	366.89	367.05
0.006	377.97	376.06	376.96	376.61	376.88	379.44	372.27	368.44	368.28	371.20	371.40	370.77
0.007	380.13	378.38	379.03	378.98	379.15	379.44	374.78	372.44	371.54	370.82	370.45	371.76
0.008	383.48	380.17	380.51	380.63	380.56	381.29	378.04	372.69	372.06	373.31	376.35	374.45
0.009	386.35	383.02	383.50	382.73	383.29	382.84	377.58	374.74	371.66	375.88	375.09	372.29
0.01	388.82	384.62	385.57	383.80	384.95	386.39	381.84	377.80	378.17	377.29	378.48	374.09
L	2.962	2.983	2.9925	2.997	2.999	3	2.974	2.995	3.005	3.010	3.012	3.012

Table 8. SDRL of the Shewhart chart with MAD outlier detection for standard normal and t ($v = 100$) distributions.

SDRL	Standard Normal Distribution						T-Distribution $v = 100$					
α/m	25	50	100	250	500	1000	25	50	100	250	500	1000
0	596.32	468.02	414.04	383.16	374.68	368.66	607.33	467.10	411.17	377.81	369.81	360.34
0.001	600.30	472.43	417.55	385.96	377.25	372.35	612.79	467.20	406.53	380.03	366.57	366.34
0.002	607.28	476.91	419.54	388.33	379.39	371.64	627.84	471.38	410.10	376.44	372.50	365.67
0.003	615.99	477.95	422.41	389.95	380.10	375.45	617.88	474.03	415.42	378.88	374.34	366.23
0.004	615.39	481.58	423.16	392.96	382.24	377.77	636.02	476.02	416.09	385.14	377.04	371.24
0.005	625.95	486.00	427.29	393.54	385.71	380.37	650.31	480.25	420.69	387.36	377.24	370.45
0.006	636.60	488.82	430.22	396.23	386.61	383.77	637.39	483.89	423.95	390.68	383.17	376.49
0.007	646.05	490.89	433.11	399.86	389.30	384.98	630.84	488.73	429.10	390.52	381.42	377.88
0.008	656.27	496.12	435.44	400.84	391.44	385.54	654.64	489.53	430.19	391.87	386.60	378.02
0.009	654.98	500.43	439.24	403.24	393.19	387.01	675.27	499.78	423.47	395.23	385.19	380.11
0.01	669.83	502.93	441.89	404.61	395.16	390.51	672.71	493.77	436.38	399.23	390.17	380.26
L	2.962	2.983	2.9925	2.997	2.999	3	2.974	2.995	3.005	3.010	3.012	3.012

For better visuals of the results, we depict the ARL results (Tables 3, 5 and 7) in Figures 2 and 3 and the SDRL results (Tables 4, 6 and 8) in Figures 4 and 5.

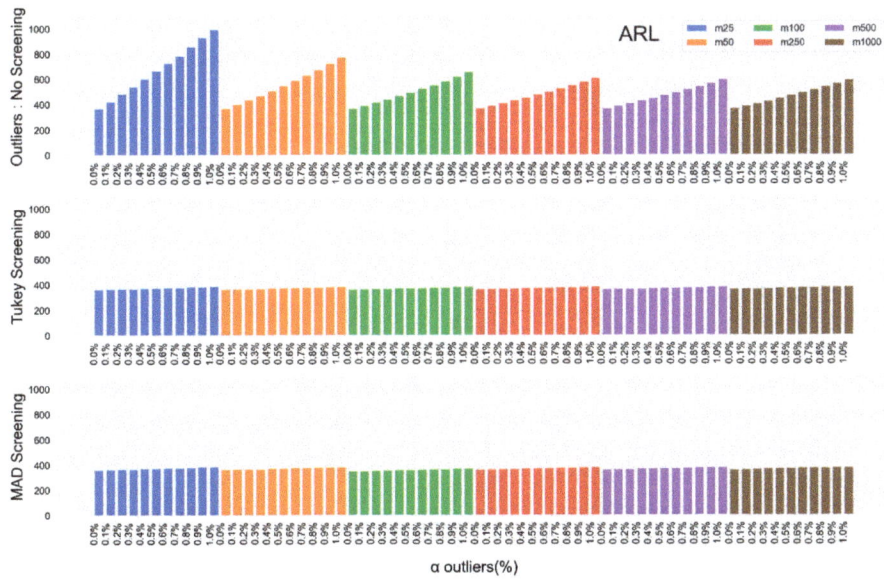

Figure 2. In-control ARL values for the Shewhart chart from standard normal distribution in the presence of outliers with and without outlier screening.

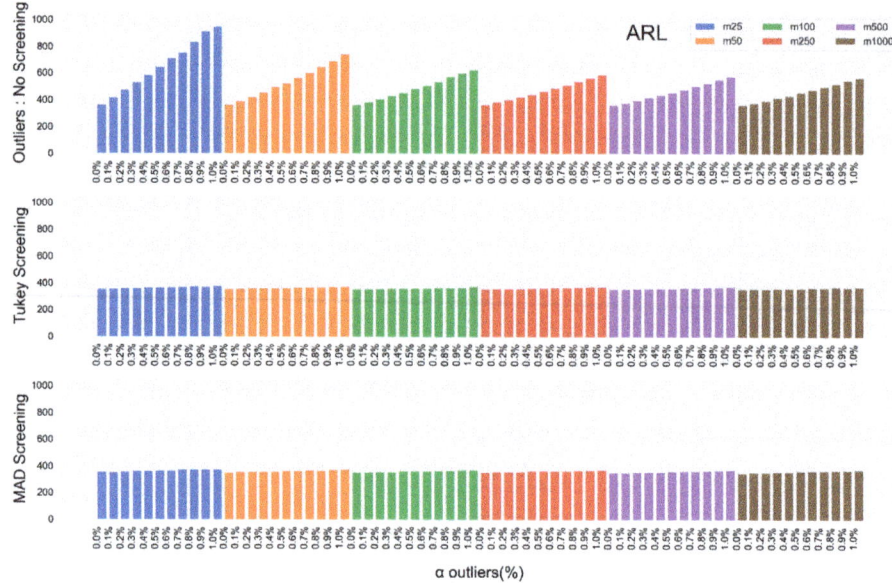

Figure 3. In-control ARL values for the Shewhart chart from t-($v = 100$) distribution in the presence of outliers with and without outlier screening.

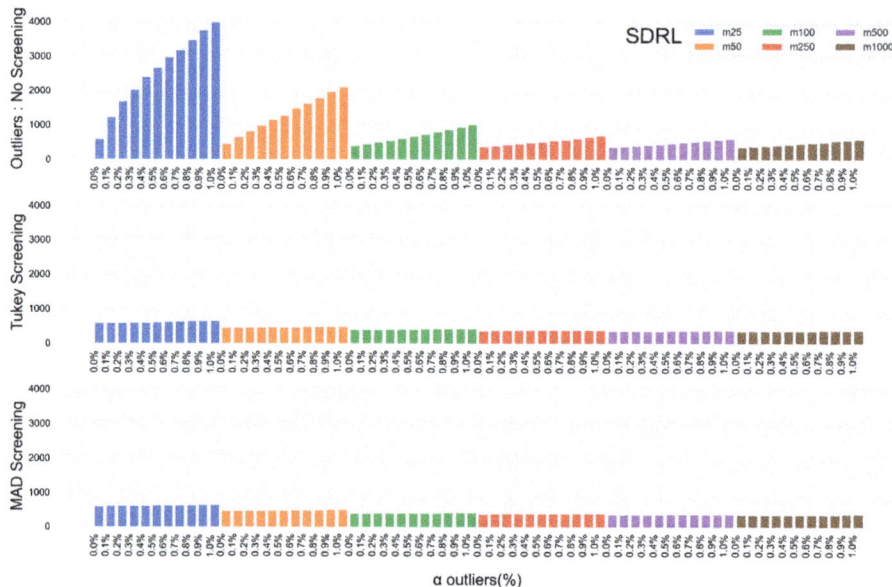

Figure 4. In-control SDRL values for the Shewhart chart from standard normal distribution in the presence of outliers with and without outlier screening.

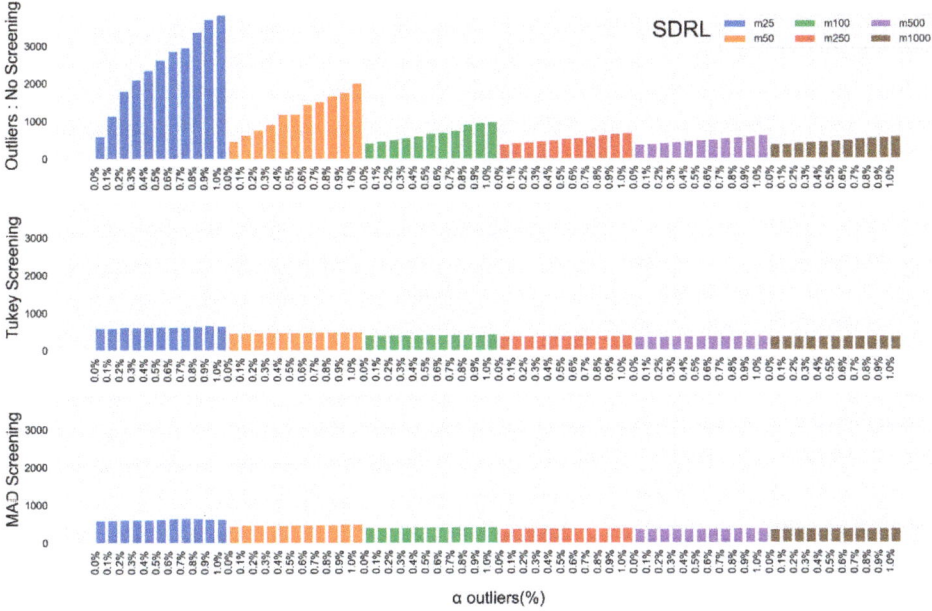

Figure 5. In-control SDRL values for the Shewhart chart from t-(v = 100) distribution in the presence of outliers with and without outlier screening.

4. Discussion

We summarize the findings of the study under the following subsections: (a) parameter estimation effect on the Shewhart control chart, (b) effect of outliers on Shewhart $\overline{X}$ chart performance, and (c) improvement of outliers screening models on the Shewhart $\overline{X}$ chart performance. Through the discussion, we use the run length properties as a yardstick for measuring the performance of the charts.

4.1. Parameter Estimation Effect on the Shewhart $\overline{X}$ Control Chart

Theoretically, when the Shewhart charts parameters are known, the limit L corresponding to the IC $ARL_0 = 370$ is $L = 3$. When the parameters are estimated from phase-I samples, the first effect of the estimation is the change in L. The control limit L deviates from its theoretical value as much as the sample size m reduces. That implies, the smaller the sample size m, the farther the control limit from the theoretical value. This is noticeable in Tables 1 and 2, as L changes as the sample size does. We compute Ls based on 100,000 iterations of simulation. Secondly, in the introduction of shifts, which makes the process OC, the RL properties values of the estimated parameters are bigger than the theoretical values. This indicates that the chart with estimated parameters are slower in detecting shifts in the process as compared to the chart with known parameters. For instance, (cf. Tables 1 and 2), with $m = 1000$, $\delta = 0.5$ the resulting ARL_1 and $SDRL_1$ are 156.42 and 158.84 for normal distribution and 150.92 and 160.46 for t-distribution respectively. However, with $m = 25$, $\delta = 0.5$ ARL_1 and $SDRL_1$ are 190.12 and 333.88 for normal distribution and 194.06 and 340.19 for t-distribution respectively.

4.2. Effect of Outliers on Shewhart $\overline{X}$ Control Chart performance

Haven noticed the effect of parameter estimation on Shewhart chart performance; one major cause could be the presence of outliers in the dataset. The results in Tables 3 and 4 prove that extreme values in the sample causes great havoc to the performance of the process. As discussed earlier in Section 4, $\alpha = 0$ indicates absence of outliers, and the presence of outliers if otherwise. We observe jumps in the values of IC ARL and SDRL from Tables 3 and 4. With different combinations of α and m, we say the bigger the value of α and the smaller the value of m, the gross the effect of the outliers on the chart. Take for instance, in the normal environment, the ARL and SDRL values of just 1% of outliers ($\alpha = 0.01$) for when $m = 1000$ as against when $m = 25$. It shocks to see the ARL and SDRL jumped from 592.55 and 612.00 to 996.3 and 4012.72 respectively. However, in the t-distribution, ARL and SDRL values of 1% of outliers ($\alpha = 0.01$) for when $m = 1000$ as against when $m = 25$, are 575.08 and 591.55 to 953.61 and 3823.24 respectively.

4.3. Improvement of Outliers Screening Models on Shewhart Chart Performance

The proposed remedy for the effect of outliers on the Shewhart chart works perfectly. The incorporation of Tukey and MAD outlier-screening models in the Shewhart chart normalizes the outlier effects and restores the performance even much better than it was. To access the effect of these two screening methods, we present Figures 2–5, displaying the IC ARL values with $m = 25, 50, 100, 250, 500$ and 1000, and the magnitude $w = 3$, without outliers screening, alongside the IC ARL whose outliers are screened with the Tukey and MAD-based models. The IC ARL that are supposed to be around the target 370 has jumped to more than 250% increment due to the effect of outliers. However, with our proposed screening models; both Tukey and MAD-based models; the IC ARL is returned back to its target with less than 5% increment and decrement. The IC SDRL also exhibits the same pattern; in fact, its improvement is more appreciable as compared to the ARL's.

5. Illustrative Example

In the manufacturing industry, semiconductor lithography (photolithography) refers to the formation of three-dimensional images on the substrate for subsequent transfer of the pattern to the substrate. A keynote aspect of this process is the bake process, both the pre (soft)-bake and post (hard)-bake. In this section, we implement the Shewhart chart with the proposed outlier detection models on the flow width measurement of a hard bake process. In the subsequent subsections, we give a brief overview of the hard-bake process and then application of the Shewhart chart on the dataset extracted from such a process (the Basics of Microlithography n.d.).

5.1. The Post (Hard) Bake Process

A typical photolithography process consist of the following sequence of operation: substrate preparation, photoresist spin coat, pre-bake, exposure, post-exposure bake, development and finally the post-bake. The hard-bake process, as the name implies, is used to harden the final resist image so that it will withstand the harsh environments of etching. This post-bake ensures complete removal of solvent, improving adhesion in wet etch processes and resistance to plasma etches. Practitioners use different temperatures depending on the material under study. However, the temperature should be carefully chosen and not more than 200 °C. A major characteristic of this process is the wafer. Recall that the word lithography is a combination of two Greek words: lithos meaning stones and graphia, meaning to write. Our stones in this case are silicon wafers and the patterns are written with photoresist, which are sensitive polymers. Figures 6 and 7 depict a typical photolithography flowchart and the hard-bake process.

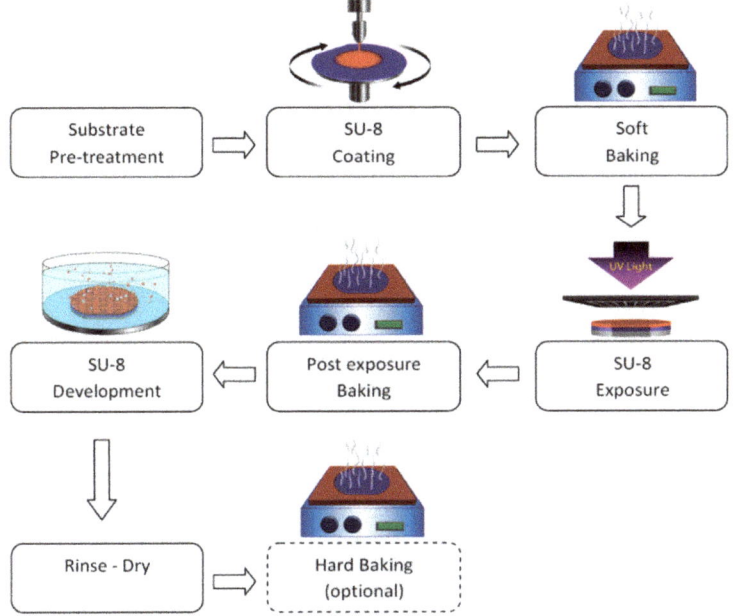

Figure 6. A flowchart of a photolithography process of semiconductor manufacturing industry.

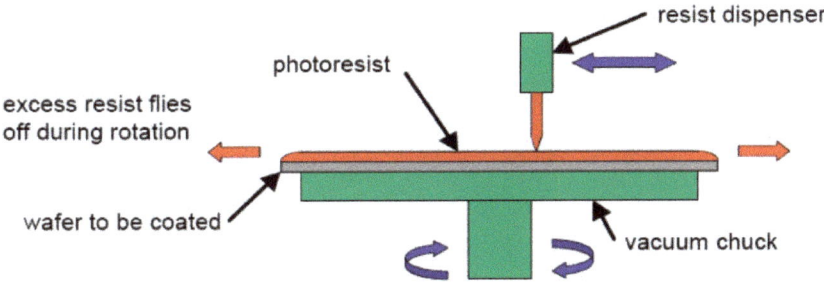

Figure 7. Illustration of hard-bake process.

5.2. Application of Shewhart Control Charts with Outlier

In this section, we implement the findings of this study on a set of data generated from a semiconductor manufacturing of a hard-bake process, which monitors the flow width measurement of wafers [1]. The variable of interest is the flow width measurement (in microns) for the hard-brake process. The data consist of 25 IC phase-I samples and 10 phase-II samples each of sample size 5. The process mean and standard deviation of the phase-I samples are 16.7163 and 3.5167, respectively. Therefore, we use these estimates to setup Shewhart chart control limits for monitoring phase-II samples. Figure 8 shows all phase-I sample points staying within the limits and 3 of the phase-II sample points stretching beyond the LCL making them OoC due to some assignable cause of variation.

Prior to setting the limits, we test the data for possible autocorrelation. The data is autocorrelation-free as the Durbin–Watson (DW) test result proves. The value of the DW test statistics is DW = 1.7564 and the critical values at 1% level of significance are $d_L = 1.19$, and $d_U = 1.31$. By the interpretation explained in Table 9, we fail to reject the null hypothesis and conclude that there is no evidence of autocorrelation in the data.

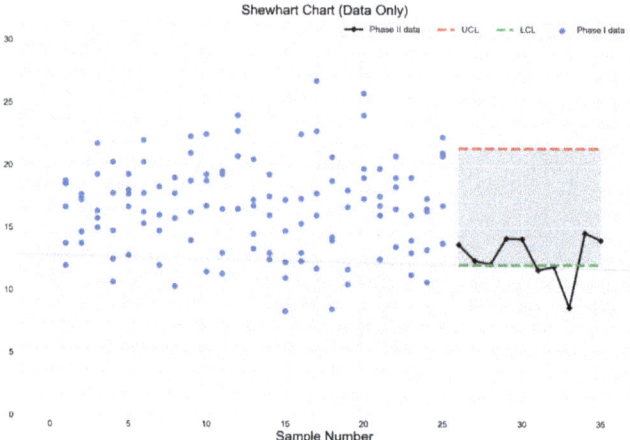

Figure 8. Scatter plot of phase-I sample and the Shewhart chart with estimated parameters.

Table 9. Interpretation of Durbin–Watson autocorrelation test.

Categories	Decision Rules	Decisions
$0 \leftrightarrow d_L = 1.19$	Reject H_0 : positive autocorrelation	
$d_L = 1.19 \leftrightarrow d_U = 1.31$	Inconclusive	
$d_U = 1.31\ 4 \leftrightarrow d_L = 2.69$	Do not reject H_0 : no autocorrelation	$1.31 < \text{DW} = 1.7564 < 2.69$
$4 - d_L = 2.69 \leftrightarrow 4 - d_U = 2.81$	Inconclusive	
$4 - d_U = 2.81 \leftrightarrow 4$	Reject H_0 : negative autocorrelation	

Furthermore, we introduce a 5% of outliers to the phase-I samples, to illustrate the argument that the presence of outliers affects the performance of control charts. This subsequently increased the mean and standard deviation by 4% and 25% respectively resulting to an increased UCL and decreased LCL. The changes in the control limits implies a wider range of the boundaries. Therefore the resulting control charts is less efficient as compared to the previous one without outliers. Figure 9 depicts this.

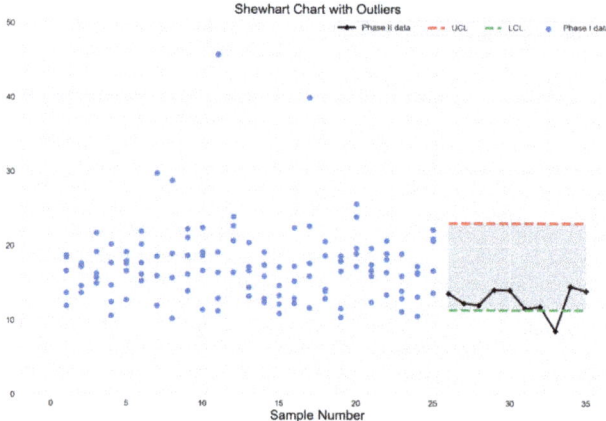

Figure 9. Scatter plot of phase-I sample and the resulting Shewhart chart with estimated parameters and 5% of outliers with magnitude 3.

5.3. Application of Shewhart Outlier Detection Model

Having established the deficiency of the Shewhart chart with outliers on the dataset; we employ our proposed outlier detection model with the Shewhart chart explained in Section 2.4 to rectify this shortcoming. Figure 10 shows the application of the Shewhart Tukey-based model. It is evident there in that the chart was not only able to restore the efficiency of the chart as there were no outliers, detecting 3 OoC sample points, but also to identify the outliers in the phase-I sample points. Similarly, Figure 11 portrays the scenario when the Shewhart MAD-based model is applied on the monitoring stage. Despite the presence of outlier in the dataset, the chart is able to detect the OC sample points as much as it does when there were no outliers.

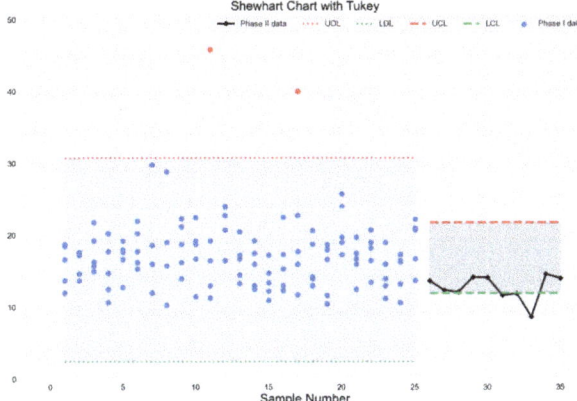

Figure 10. Scatter plot of phase-I sample and the resulting Shewhart chart with Tukey outlier detection screening.

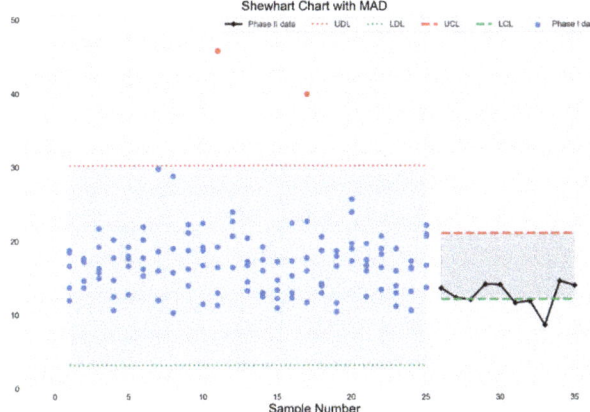

Figure 11. Scatter plot of phase-I sample and the resulting Shewhart chart with MAD-outlier detection screening.

6. Conclusions

In this article, we evaluate the performance of the Shewhart control chart for location monitoring with estimated parameters. The study substantiates the effect of estimation error and the variability in the practitioners' choice of phase-I samples on the chart, especially when the samples are prone to outliers. Increasing the phase-I sample size (although not practicably) will to some extent reduce the gross impact on the Shewhart chart. The results of this study further prove that incorporation of the non-parametric outlier screening models, Tukey and MAD, in the design of the Shewhart chart is more practicable as it requires less phase-I samples and yields better results. Another advantage of this study lies in the simplicity of its design and ease of usage. The study rounds up with an illustrative example with a photolithography real data. A comparison of the two detection models, Tukey and MAD, reveals that duo relatively efficient. The study is limited to operate within the univariate setup, while focusing on multivariate setup will be a great advantage and we plan a future study for that. Also, proposed charts are memory-less, which implies they are suitable for monitoring large shift. However, the idea of the study is not only applicable in Shewhart multivariate setup, but also extendable to other control charts, like exponentially weighted moving average and cumulative sum charts both univariate and multivariate setups.

Author Contributions: Conceptualization, I.A.R. and N.A.; Data curation, I.A.R. and M.R.A.; Investigation, M.H.L.; Methodology, N.A.; Resources, M.H.L.; Supervision, M.R. and M.R.A.; Writing—original draft, I.A.R.; Writing—review & editing, M.H.L., M.R. and N.A. All authors have read and agreed to the published version of the manuscript.

Funding: This study was partially funded by the Ministry of Higher Education, Malaysia through the Industry-International Incentive Grant for Universiti Teknologi Malaysia (grant number 01M30).

Conflicts of Interest: The authors declare no conflict of interest.

References

1. Montgomery, D.C. *Introduction to Statistical Quality Control*, 6th ed.; John Wiley & Sons Inc.: Hoboken, NJ, USA, 2009.
2. Chen, G. The mean and standard deviation of the run length distribution of X charts when control limits are estimated. *Stat. Sin.* **1997**, *7*, 789–798.
3. Jensen, W.A.; Jones-Farmer, L.A.; Champ, C.W.; Woodall, W.H. Effects of parameter estimation on control chart properties: A literature review. *J. Qual. Technol.* **2006**, *38*, 349–364. [CrossRef]
4. Psarakis, S.; Vyniou, A.K.; Castagliola, P. Some Recent Developments on the Effects of Parameter Estimation on Control Charts. *Qual. Reliab. Eng. Int.* **2014**, *30*, 1113–1129. [CrossRef]
5. Saleh, N.A.; Mahmoud, M.A.; Keefe, M.J.; Woodall, W.H. The difficulty in designing shewhart X and X control charts with estimated parameters. *J. Qual. Technol.* **2015**, *47*, 127–138. [CrossRef]
6. Goedhart, R.; Da Silva, M.M.; Schoonhoven, M.; Epprecht, E.K.; Chakraborti, S.; DoesA, R.J.M.M. Veiga Shewhart control charts for dispersion adjusted for parameter estimation. *IISE Trans.* **2017**, *49*, 838–848. [CrossRef]
7. Dasdemir, E.; Weiß, C.; Testik, M.C.; Knoth, S. Evaluation of Phase I analysis scenarios on Phase II performance of control charts for autocorrelated observations. *Qual. Eng.* **2016**, *28*, 293–304. [CrossRef]
8. Anscombe, F.J. Rejection of Outliers. *Technometrics* **1960**, *2*, 123–146. [CrossRef]
9. Grubbs, F.E.; Beck, G. Extension of Sample Sizes and Percentage Points for Significance Tests of Outlying Observations. *Technometrics* **1972**, *14*, 847–854. [CrossRef]
10. Hawkins, D.M. *Identification of Outliers*; Chapman and Hall: London, UK, 1980.
11. Beckman, R.J.; Cook, R.D. Outlier . . . . s. *Technometrics* **1983**, *25*, 119–149.
12. Barnett, V.; Lewis, T. Outliers in Statistical Data. 3rd edition. J. Wiley & Sons 1994, XVII. 582 pp., £49.95. *Biom. J.* **1995**, *37*, 256.
13. Torng, C.-C.; Lee, P.-H. ARL Performance of the Tukey's Control Chart. *Commun. Stat. Simul. Comput.* **2008**, *37*, 1904–1913. [CrossRef]
14. Tercero-Gomez, V.; Ramirez-Galindo, J.; Cordero-Franco, A.; Smith, M.; Beruvides, M. Modified Tukey's Control Chart. *Commun. Stat. Simul. Comput.* **2012**, *41*, 1566–1579. [CrossRef]
15. Fan, S.-K.S.; Huang, H.-K.; Chang, Y.-J. Robust Multivariate Control Chart for Outlier Detection Using Hierarchical Cluster Tree in SW2. *Qual. Reliab. Eng. Int.* **2013**, *29*, 971–985. [CrossRef]
16. Khaliq, Q.-U.-A.; Riaz, M. Robust Tukey-CUSUM Control Chart for Process Monitoring. *Qual. Reliab. Eng. Int.* **2016**, *32*, 933–948. [CrossRef]

17. Fitriyah, H.; Budi, A.S. Outlier Detection in Object Counting based on Hue and Distance Transform using Median Absolute Deviation (MAD). In Proceedings of the 2019 4th International Conference on Sustainable Information Engineering and Technology, Lombok, West Nusa Tenggara, Indonesia, 28–30 September 2019; pp. 217–222.
18. Abbas, N. A robust S^2 control chart with Tukey's and MAD outlier detectors. *Qual. Reliab. Eng. Int.* **2020**, *36*, 403–413. [CrossRef]
19. Abbas, N.; Abujiya, M.R.; Riaz, M.; Mahmood, T. Cumulative Sum Chart Modeled under the Presence of Outliers. *Mathematics* **2020**, *8*, 269. [CrossRef]
20. Dequeant, K.; Vialletelle, P.; Lemaire, P.; Espinouse, M.L. A literature review on variability in semiconductor manufacturing: The next forward leap to Industry 4.0. In Proceedings of the Winter Simulation, Arlington, VA, USA, 11–14 December 2016; pp. 2598–2609.
21. Spanos, C.J. Statistical Process Control in Semiconductor Manufacturing. *Proc. IEEE* **1992**, *80*, 819–830. [CrossRef]
22. Huh, I. Multivariate EWMA Control Chart and Application to a Semiconductor Manufacturing Process. Ph.D. Thesis, Department of Statistics, McMaster University, Hamilton, ON, Canada, 2010.
23. Baud-Lavigne, B.; Bassetto, S.; Penz, B. A broader view of the economic design of the X-bar chart in the semiconductor industry. *Int. J. Prod. Res.* **2010**, *48*, 5843–5857. [CrossRef]
24. Higashide, M.; Nishina, K.; Kawamura, H.; Ishii, N. Statistical process control for semiconductor manufacturing processes. *Front. Stat. Qual. Control* **2010**, *9*, 71–84.
25. Quesenberry, C.P. The Effect of Sample Size on Estimated Limits for $\overline{X}$ and X Control Charts. *J. Qual. Technol.* **1993**, *25*, 237–247. [CrossRef]
26. Albers, W.; Kallenberg, W.C.M. Estimation in Shewhart control charts: Effects and corrections. *Metrika* **2004**, *59*, 207–234. [CrossRef]
27. Dovoedo, Y.H.; Chakraborti, S. Effects of Parameter Estimation on the Multivariate Distribution-free Phase II Sign EWMA Chart. *Qual. Reliab. Eng. Int.* **2017**, *33*, 431–449. [CrossRef]
28. Maleki, M.R.; Castagliola, P.; Amiri, A.; Khoo, M.B.C. The effect of parameter estimation on phase II monitoring of poisson regression profiles. *Commun. Stat. Simul. Comput.* **2019**, *48*, 1964–1978. [CrossRef]
29. Jardim, F.S.; Chakraborti, S.; Epprecht, E.K. Two perspectives for designing a phase II control chart with estimated parameters: The case of the Shewhart $\overline{X}$ Chart. *J. Qual. Technol.* **2020**, *52*, 198–217. [CrossRef]

© 2020 by the authors. Licensee MDPI, Basel, Switzerland. This article is an open access article distributed under the terms and conditions of the Creative Commons Attribution (CC BY) license (http://creativecommons.org/licenses/by/4.0/).

Article

A Functional Data Analysis for Assessing the Impact of a Retrofitting in the Energy Performance of a Building

Miguel Martínez Comesaña [1,*], Sandra Martínez Mariño [1], Pablo Eguía Oller [1], Enrique Granada Álvarez [1] and Aitor Erkoreka González [2]

[1] Department of Mechanical Engineering, Heat Engines and Fluid Mechanics, Industrial Engineering School, University of Vigo, Maxwell s/n, 36310 Vigo, Spain; samartinez@uvigo.es (S.M.M.); peguia@uvigo.es (P.E.O.); egranada@uvigo.es (E.G.Á.)
[2] ENEDI Research Group, Department of Thermal Engineering, University of the Basque Country, 48013 Bilbao, Spain; aitor.erkoreka@ehu.eus
* Correspondence: migmartinez@uvigo.es

Received: 13 February 2020; Accepted: 19 March 2020; Published: 7 April 2020

Abstract: There is an increasing interest in reducing the energy consumption in buildings and in improving their energy efficiency. Building retrofitting is the employed solution for enhancing the energy efficiency in existing buildings. However, the actual performance after retrofitting should be analysed to check the effectiveness of the energy conservation measures. The aim of this work was to detect and to quantify the impact that a retrofitting had in the electrical consumption, heating demands, lighting and temperatures of a building located in the north of Spain. The methodology employed is the application of Functional Data Analyses (FDA) in comparison with classic mathematical techniques such as the Analysis of Variance (ANOVA). The methods that are commonly used for assessing building refurbishment are based on vectorial approaches. The novelty of this work is the application of FDA for assessing the energy performance of renovated buildings. The study proves that more accurate and realistic results are obtained working with correlated datasets than with independently distributed observations of classical methods. Moreover, the electrical savings reached values of more than 70% and the heating demands were reduced more than 15% for all floors in the building.

Keywords: retrofitting; refurbishment; functional data analysis; vectorial analysis; energy efficiency

---

1. Introduction

The building sector is considered the largest energy consumer in the European Union, representing 40% of the final energy consumption [1]. Globally, the energy consumption of this sector accounts for 20% of the total delivered energy [2]. Thus, there is increasing interest in improving the energy efficiency of buildings [3–5]. Furthermore, the potential of saving energy by renovation in Europe is considerable as two-thirds of European buildings were constructed before 1980 [6]. Building retrofitting can contribute to reduce the energy consumption of existing buildings with lower energy efficiencies. In this context, it is important to develop methodologies that can evaluate the actual impact of refurbishment on renovated buildings in terms of energy consumption, thermal comfort and lighting.

Therefore, building retrofitting is essential to prove the effectiveness of the applied energy conservation measures to check if the energy efficiency of the building has certainly improved. However, there are very few studies that actually evaluate the retrofit of buildings [7]. Most of the studies evaluate measures for building refurbishment based on energy simulation [8–10], mathematical models [11], artificial neural networks (ANN) [12] and building information modelling (BIM) [13]. Thus, most studies analyse the energy conservation measures based on model outputs and not proving the real effect on the building with monitored data. There is a performance gap between simulated and measured energy consumption. Some studies have found that the calculated heating energy consumption levels in the design phase were much lower than the measured values [14].

Some authors have evaluated the effect of refurbishment and renovations made in buildings with real data. Ardente et al. [15] presented the results of an energy and environmental assessment of retrofit actions implemented in six public buildings by using life cycle analysis (LCA). It is a common approach to evaluate the decrease of energy consumption, operational cost and environmental impact in building retrofitting [16–18]. Another approach for energy diagnosing of existing buildings is U-value in-situ measurement [19] that characterises the heat losses through the building envelope and that can be used to evaluate retrofitting actions [20]. Zavadskas et al. [21] proposed an approach to assess indoor environmental conditions before and after retrofitting of dwellings with multiplicative optimality criteria and experimental data. Hamburg et al. [14] analysed how well the energy performance targets of building refurbishment are reached by collecting energy consumption and indoor measurements after renovation and constructing simulations.

All of the previous presented methodologies to evaluate the retrofitting made in buildings use vector-based data approaches. The presented methodology in this work for evaluating the impact of a building retrofitting is based on functional analysis of monitored data before and after retrofitting. By comparison of vector and functional analysis, we demonstrate that functional analysis provides more realistic and accurate evaluations of the studied variables.

The methods for assessing building renovation have been applying vectorial analysis to the data. These methods do not take into account the observations within a day as a set when evaluating the daily behaviour of the data. As a consequence, the correlation between observations is missed. In this context, Functional Data Analysis (FDA) can be useful because it is able to detect days that do not have individual outliers, but may be far from the mean behaviour [22–26]. A proof of its application is that FDA has expanded to a great number of scientific fields related with continuous-time monitoring processes such as the environment [24,26–29], health and medical research [30,31], industrial processes [32,33], sensor technology [34,35] or even econometrics [36]. Moreover, it has also been applied with machine learning techniques in optimisation and classification problems [37,38]. Nowadays, FDA continues to expand its applications in more fields such as quality control or sports [39,40]

In this work, we propose the use of FDA for assessing the impact that building retrofitting has in the energy performance, indoor temperatures and lighting conditions of a building. The methodology was applied to a case study of the renovated building of the Rectorate of the University of the Basque Country (Spain). The novelty of this work is the application of FDA to statistically contrast the differences in the energy performance of a building before and after a retrofitting. The literature review shows that just few studies actually evaluated the retrofit of buildings with monitored data and functional analysis was not used for this application.

The samples are composed of daily curves of variables such as electricity, heating demands and temperatures. FDA allows making the contrast between samples taking into account the average behaviour of the group throughout all day [24,37,41], which would not be possible with a vectorial approach. In vectorial analysis, the data of a whole day have to be summarised in a single value to work with daily observations. To classify a day as outlier, it has to move away from the sample mean, in this case calculated with simplified daily observations [42,43].

On the one hand, to carry out the functional analysis, a functional ANOVA (FANOVA) was used to evaluate whether there are differences between monitored data in the building before and after retrofitting. On the other hand, a classical analysis of variance (ANOVA) was also used to study the differences between the samples before and after the retrofitting [43–46]. To complement the vectorial analysis, Kruskal's non-parametric test was applied to contrast if the two samples come from the same initial distribution [47–49]. In addition, some variables such as the rate of change in sample variance or the functional $\mathcal{L}_2$ distance between curves are also presented to measure the impact of the refurbishment. The vectorial method is based on the differences between the medians, and the functional method consists of measuring the distance between the curves that represent the mean functions [50–52].

The results show that FDA efficiently demonstrates that the heating demands in the building were reduced thanks to the envelope insulation, although ventilation was increased, indoor temperatures were increased and internal lighting loads were reduced. The results also show that a significant reduction in lighting consumption was achieved with the installation of LED lighting. Moreover, it is demonstrated that taking into account the correlation of the data from a functional approach is a more realistic and informative way to study how different two or more samples are.

2. Materials and Methods

2.1. Functional Data Analysis (FDA)

Functional Data Analysis (FDA) studies observations which form functions defined over a determined set T. The infinite-dimensional structure of the data enlarges the possibilities of research [23,27,53]. A random variable $\mathcal{X}$ is defined such a functional variable if it takes values in a complete metric or semi-metric space and is observed in a discrete set of points $\{t_j\}_{j=1}^{n_p} \in [a,b]$ (not necessarily equispaced) for each of the n individuals studied [54,55]. Thus, the data consist of a $\mathbf{X}$ matrix with n rows representing the different individuals and n_p columns representing the different discrete points where the functions are evaluated [55].

The functional model, through a process known as smoothing, converts the initial discrete values into a set of continuous functions over time $x(t) \in \mathcal{X} \subset \mathcal{F}$, being $\mathcal{F}$ a functional space. To estimate these functions, $\mathcal{F}$ is $\mathcal{F} = span\{\phi_1, ..., \phi_{n_b}\}$, where ϕ_k is a base function and n_b the number of basis functions necessary to build a functional sample. Although there are other types, the basis functions used commonly are spline or Fourier functions [56], and the expansion considered is [24,25,27,57]:

$$x(t) = \sum_{k=1}^{n_b} c_k \phi_k(t) \quad (1)$$

where $\{c_k\}_{k=1}^{n_b}$ represent the coefficients that shape the function $x(t)$ with respect to the chosen set of basis functions. In this way, the smoothing process consist on solving the following regularisation problem [24,25,27,57]:

$$\min_{x \in F} \sum_{j=1}^{n_p} \{z_j - x(t_j)\}^2 + \lambda \Gamma(x) \quad (2)$$

where $z_j = x(t_j) + \epsilon_j$ (being ϵ_j a value of the zero-mean random noise) is the result of observing x at the point t_j, λ a regularisation parameter that controls the intensity of the regularisation, and Γ an operator that penalises the complexity of the solution. Taking into account the expansion, Equations (1) and (2) can be expressed as [24,25,27,57]:

$$\min_{\mathbf{c}} \{(\mathbf{z} - \mathbf{\Phi c})^T (\mathbf{z} - \mathbf{\Phi c}) + \lambda \mathbf{c}^T \mathbf{R c}\} \quad (3)$$

where $\mathbf{z} = (z_1, ..., z_{n_p})^T$ is the observation vector, $\mathbf{c} = (c_1, ..., c_{n_b})^T$ the vector coefficients of the functional expansion, $\mathbf{\Phi}$ the $n_p \times n_b$ matrix of $\Phi_{jk} = \phi_k(t_j)$ elements, and $\mathbf{R}$ the matrix formed by $n_b \times n_b$ elements [24,25,57,58]:

$$R_{kl} = \langle D^2 \phi_k, D^2 \phi_l \rangle_{\mathcal{L}_2(1)} = \int_T D^2 \phi_k(t) D^2 \phi_l(t) dt \quad (4)$$

where $D^n \phi_k(t)$ represents the nth-order differential operator of the function ϕ_k. After this, it is easy to know that the solution can be calculated as follows:

$$c = (\mathbf{\Phi}^t\mathbf{\Phi} + \lambda \mathbf{R})^{-1}\mathbf{\Phi}^T\mathbf{z} \tag{5}$$

2.1.1. Functional Depths

Initially, the depth concept appeared in the multivariate statistics for measuring the centrality of a point $x \in \mathbb{R}^d$ within a specific dataset: giving greater value to the points near the center [22]. Then, some authors extended this measured to FDA [59,60]. The functional depth give us a centrality measure of a specific curve x_i with respect a set of curves $x_1, ..., x_n$ that comes from a stochastic process $\mathcal{X}(\cdot)$ in a defined interval $[a,b] \in \mathbb{R}$.

There are several functional depths in the statistical literature, but the three main are: *Fraiman–Muniz* [59], *h-modal* [61] and *Random Projections* [60]. The most used is the *h-modal* depth because it has a better frequency of correct detection than the others [22]. The *h-modal* depth defines the functional mode as the curve most densely surrounded by other curves of the dataset. In this manner, the functional depth of a curve x_i with respect the other curves in the sample is given by:

$$MD_n(x_i, h) = \sum_{k=1}^{n} K\left(\frac{||x_i - x_k||}{h}\right) \tag{6}$$

with $||\cdot||$ being a norm in the functional space, $K: \mathbb{R}^+ \to \mathbb{R}^+$ a kernel function, and h a bandwidth parameter [61]. Thus, the curve that gets the maximum value in Equation (6) is considered the functional mode. Moreover, some authors [60,61] recommended the use of $\mathcal{L}^2(l)$ norms and a truncated Gaussian kernel:

$$||x_i - x_k||_2 = \left(\int_a^b (x_i(t) - x_k(t))^2 dt\right)^{1/2} \qquad K(t) = \frac{2}{\sqrt{2\pi}} exp\left(-\frac{t^2}{2}\right), t > 0 \tag{7}$$

The principal aim of functional depths, viewed as functional dispersion measure, is the detection of outliers. As in the classical analysis, detecting and examining these curves is important because they may bias our functional estimations and because it allows us to discover the reasons that make these curves deviate from the mean. Furthermore, from a functional approach, it is essential because it may occur that the individual values of a curve are not outliers vectorially, but, instead, the complete curve is a functional outlier [22,58]. If we assume that every curve in the data come from the same stochastic process, a curve would be considered such an outlier for two reasons: it is at a significant distance from the expected function of the stochastic process or its shape represents a very different behaviour from the other curves. Therefore, the curves with functional depth below a specific C value would be considered atypical and would be removed from the sample (see [23–26]). On the other hand, it would be convenient to choose a C that provides a controlled type I error level. It should be a value that, in absence of outliers, the probability of mislabelling a correct data as outlier would be approximately a 1% [23–26]:

$$P(D_n(x_i) \leq C) = 0.01, \quad i = 1, ..., n \tag{8}$$

In this way, the chosen C will be first percentile of the depths distribution chosen. Since this distribution is unknown this, percentile must be estimated using the sample data. For this purpose, there are two different bootstrap techniques: trimming bootstrap [62] and weighting bootstrap [63]. Some studies demonstrate that, despite having a larger incorrect outlier detection, the trimming bootstrap has a better performance detecting the curves that are actually outliers [22,64].

2.1.2. Functional Test ANOVA (FANOVA)

Any test or contrast that can be made in a vectorial analysis can have a functional version that usually provides more relevant information. An example of this is the classical ANOVA. Its functional version, although also contrasting the mean levels of a variable, is based on k independent samples $X_{ij}(t)$, $j = 1, ..., n_i$ $t \in [a,b]$ drawn from $\mathcal{L}^2(l)$ processes X_i, $i = 1, ..., k$ such that $E(X_i(t)) = m_i(t)$ [65–68]. If we have a functional sample classified in several groups such as $\{\mathcal{X}_i, \mathcal{G}_i\}_{i=1}^n \in \mathcal{F} \times \mathcal{G} = \{1, ..., G\}$, where $\mathcal{G}$ is a discrete variable that tell us the membership group, the contrast will be:

$$\begin{cases} H_0 : \overline{X}_1 = \overline{X}_2 = ... = \overline{X}_G \\ H_1 : \exists k, j \text{ s.t. } \overline{X}_k \neq \overline{X}_j \end{cases} \qquad (9)$$

After a few operations, as shown in [50,51], it is possible to go from classic (F_n) to functional statistic (V_n).

$$F_n = \frac{\sum_i^G n_{i.}(\overline{Y}_{i.} - \overline{Y})^2/(G-1)}{\sum_i^G \sum_j^{n_i}(Y_{ij} - \overline{Y}_{i.})^2/(n-G)} \implies V_n = \sum_{i<j}^{n_i} ||\overline{Y}_{i.} - \overline{Y}_{j.}||^2 \qquad (10)$$

In addition, according to Cuevas et al. [50] and Tarrío-Saavedra et al. [51], the asymptotic distribution of V_n under H_0 is the same as the following statistic:

$$V := \sum_{i<j}^{G} ||Z_i(t) - C_{ij}Z_j(t)||^2 \qquad (11)$$

where $C_{ij} = (p_i/p_j)^{1/2}$ and $Z_1(t), ..., Z_G(t)$ are independent Gaussian processes with mean 0 and covariance functions $K_i(s,t)$.

Finally, H_0 will be rejected, at a level α, whenever $V_n > V$ where $P_{H_0}\{V > V_\alpha\} = \alpha$ [52]. Because in practise it is not easy to estimate the distribution of V, usually, it is necessary to implement a Monte Carlo procedure through which we get for each $i = 1, ..., G$, N iid observations

$$Z_{il}^* = (Z_{il}^*(t_1), ..., Z_{il}^*(t_m)), \quad l = 1, ..., N \qquad (12)$$

from a m-dimensional Gaussian random variable with mean 0 and covariance matrix $(\hat{K}_i(t_p, t_q))_{1 \leq p, q \leq m}$. The functional $\mathcal{L}^2(l)$-distances $||Z_i(t) - C_{ij}Z_j(t)||^2$ are approximated by the $\mathbb{R}^m$-Euclidean distances $||Z_{il}^* - C_{ij}Z_{jl}^*(t)||^2$. Ultimately, the replications $\tilde{V}_l$ of V are

$$\tilde{V}_l = \sum_{i<j}^{G} ||Z_{il}^* - C_{ij}Z_{jl}^*|| \qquad (13)$$

and the distribution of V is approximated from the empirical distribution to a sample $\tilde{V}_1, ..., \tilde{V}_N$ [50,51].

2.1.3. Functional Strengths

FDA has numerous advantages but the following are the ones that make FDA suitable for studying daily behaviours in energy variables [24,28,58]:

- It is not mandatory to have prior information on data distribution. The study does not depend on or is not limited to certain distributions.
- The analysis takes into account time intervals as a unit. The sample analysed focusses on complete time units such as days, months or years.
- Analysis of homogeneity. The definition of outliers is different; it is based on the idea that, even though data do not surpass the cut-off, if they show constant deviations, they will be identified as outlier.

- Possibility of study trends. Besides calculating mean functions or detect outliers, it is also possible to study slight variations from the normal data behaviour of the data without outliers.
- Complete analysis of the time spectrum. Before this approach, most analyses were based on the values obtained in a given grid of discrete points. On the contrary, with FDA, it is possible to work with the entire time set in a continuous way.

2.2. Building Description

The case study of this work is the Rectorate building of the University of the Basque Country (Spain) constructed in the 1970s. It is a large building divided in three blocks (west, central and east), as shown in Figure 1. For the purpose of this work, only the west block was considered, which is an office building that includes a nursery. Four storeys form the block: floor 0 (Ground Floor) and Floor 2 (2F) consist of rooms and offices while Floor 1 (1F) and Floor 3 (3F) are mainly open spaces (see Figure 2). As the building use is office work, the occupation takes place only during weekdays, being reduced during summer. There is lighting consumption from 7 a.m. to 8 p.m., except on the second floor, where lighting consumption ends at 6 p.m. The building has a centralised heating system with hot water radiators powered by a campus district heating. The heating works the whole year following a control program according to a schedule from 6 a.m. to 7 p.m. except in July and August, when the heating remains off. A refurbishment of the building was made in the summer of 2016 on the ground floor and in the summer of 2017 in the rest of the building to reduce its energy consumption.

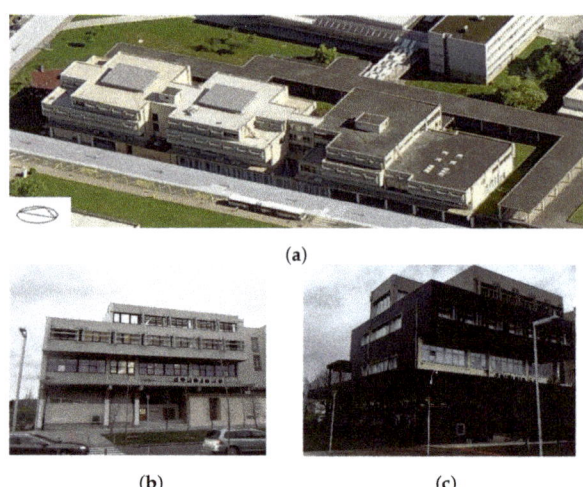

Figure 1. (a) View of the whole rectorate building that is formed by three blocks (west, central and east). This work focussed in the study of the retrofitting measures carried out in the west block. (b) West block before retrofitting. (c) West block after retrofitting.

2.2.1. Building Description before Retrofitting

Before the retrofitting, the building did not have any insulation. Most of the façade was built with precast reinforced concrete panels with non-ventilated air gap. Regarding the windows, some were single-glazed with wooden frame, and others were double-glazed but with aluminium frame without thermal break. Before retrofitting, there was no air conditioning or mechanical ventilation system.

2.2.2. Building Description after Retrofitting

The refurbishment consisted in several measures to improve the building's envelope and energy system. The façade was insulated with vacuum insulated panels (VIPs) to reduce heat losses through the envelope. Some windows were replaced by high-performance windows. To improve the lighting consumption of the building, LED lighting was installed. A ventilation system with heat recovery was added on each floor. Furthermore, thermostatic valves were installed to improve the control of the hot water radiators. A detailed description of the refurbishment and an assessment of the heat loss coefficient of the building is provided in [20].

Figure 2. Plans of the four floors in the west block of the building with their space distribution and dimensions. The use of each space is represented with colours: offices (brown), nursery (green), storage rooms (grey), corridors (pink) and server room (blue). In addition, the distribution of the sensors is also displayed: circles, illuminance, temperature and relative humidity; triangles, calorimeters; and square, lighting consumption.

2.2.3. Monitoring Description of the Building

The monitoring started before the refurbishment to investigate the necessary energy conservation measures. Thus, monitored data before and after retrofitting from sensors located all around the building are available. Monitored data before retrofitting correspond to 2016 and 2017, and data after retrofitting correspond to s 2018 and 2019; all data are minute-by-minute values. In particular, on the ground floor, the renovation started a year earlier with the insulation of a false ceiling and the installation of LED lighting. In this case, the years considered before retrofitting are 2015 and 2016. The monitored variables include outdoors conditions, indoors conditions and building heating and lighting consumption. Table 1 presents all the monitored variables in the building used to assess the impact that the retrofitting had on the building's energy performance, comfort and lighting. Indoor condition sensors are located in several points at each floor, as shown in Figure 2. The electrical consumption and the heating demand is provided per floor. Further information about the monitoring of the building can be found in the work carried out by Erkoreka et al. [69].

2.3. Pre-Processing Data

Before presenting the results, it is necessary to explain the specific smoothing process performed in this study. On the one hand, because there are parts of the day in which the variables are constant, the basis functions chosen are splines. On the other hand, to select the optimal number of basis, the determination coefficient R^2 was taken into account to measure the smoothing adjustment in relation to the raw data. As shown in Figure 3, the criterion was to select the minimum number of basis (in a given grid) where the R^2 surpasses the value of 99%.

Table 1. Monitored variables in the building to assess the impact of the refurbishment in the energy performance, illuminance and comfort of the building.

Type of Measurement		Monitored Variable	Units	Sensor
Indoor conditions		Illuminance	[LUX]	Siemens 5WG1 255-4AB12
		Temperature	[°C]	ARCUS SK04-S8-CO2-TF
Electrical Consumption	Before retrofitting	Lighting + elec. equipment	W	Power meters ABB EM/S and ABB a41/43 per floor
	After retrofitting	Lighting	W	
		Ventilation + elec. equipment	W	Power meters ABB EM/S and ABB a41/43 per floor
Heating consumption		Thermal energy of the heating water	W	Calorimeter: Kamstrup Multical 602 and ZENNER Zelsius (DN20)

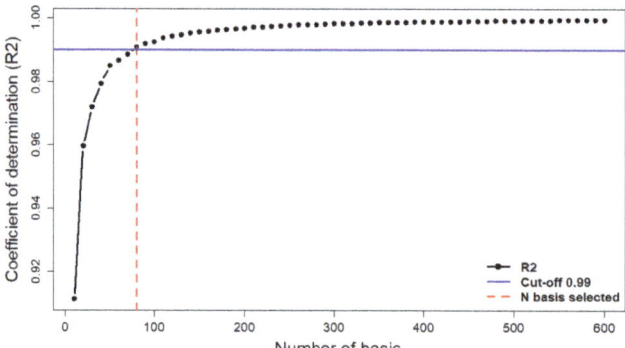

Figure 3. Example of the process of selecting the optimum number of basis for smoothing the original sample. The red line represents the minimum number of basis where the R^2 is higher than 0.99.

In addition, it is important highlight the *data cleaning* that was carried out using a functional approach. It is known that throughout all the analysed years there were many days when the building was closed and the information that these days provide is not only not useful, but can also distort the final results. To solve this problem, an algorithm that searches for these days and deletes them from the sample was developed (see Algorithm 1).

After an exploratory analysis of the data, the values of the chosen parameters for Algorithm 1 are: $\beta_1 = 500$, $\beta_2 = 250$, $\alpha = 0.25$, and $\theta = 0.5$. Moreover, because there are parts of the day in which the variables are constant, the chosen basis functions are spline. With the application of the algorithm to the data, the sample becomes smaller but only with relevant days that take into account the normal behaviour of the building. Figure 4 illustrates the performance of the algorithm. It can be seen that it is capable of deleting the days with an abnormal behaviour without affecting the bulk of the data. The days when the building was unoccupied and closed, and therefore with very small or no electrical consumption, are detected and eliminated in the picture on the right of Figure 4.

Figure 4 also presents the mean functions (in form of curves) and the change that they suffer after deleting non-representative days. Working with non-representative days will produce erroneous results in any study; for example, if an ANOVA test is performed, the test may not reject the equality of means even if the groups are different.

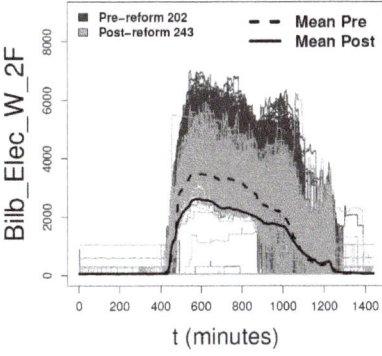 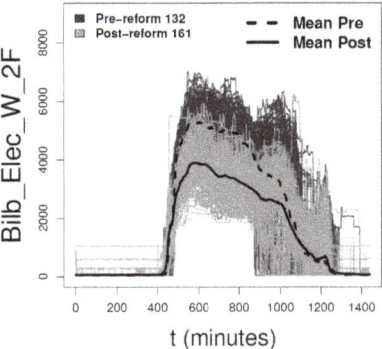

Figure 4. Performance of the *Functional cleaning* algorithm with an example variable: Electrical demand on the second floor of the building in the representative months (in dark gray the data before the refurbishment, in light gray after). The number of days taken into account in each sample is also shown: (**left**) the raw data and the mean functions separated between before and after the refurbishment; and (**right**) the data after applying the algorithm and the mean functions separated between before and after the refushbishment.

Algorithm 1: Functional cleaning.

Input: Data divided in groups and the parameters: $\beta_1, \beta_2, \alpha, \theta$.
Output: Data without inappropriate days.

1 Transform the data to funcional format: 1440 minute data each day.
2 Searching for missing values (NAs). The daily limits are:
    β_1 NAs per day β_2 consecutive NAs per day
3 Delete the days that exceeded the daily limits.
4 Approximation, with an interpolation technique, of the remaining missing values.
5 Calculate the variability of every daily curve in the sample.
6 Delete the curves that:

- Have a variability less than or equal to a percentage $\alpha \in \mathbb{R}$ of the average sample variability.
- Be below the sample mean function for at least a percentage $\theta \in \mathbb{R}$ of the day.

---

3. Results and Discussion

The effects of the refurbishment carried out in 2017 in the Rectorate building of the University of the Basque Country were analysed. In this analysis, lighting consumption, illuminance, indoor temperatures and heating demand were studied. These variables were measured every minute between 2016 and 2019 (changing 2016 for 2015 in the case of the ground floor) and only taking into account those months in which the heating systems operate significantly (October to March). As the retrofitting started in the summer of 2017, the months of this year after the summer are not suitable for analysis. In this way, data were divided into nine months before retrofitting (six months in 2016 and three months in 2017) and nine months after retrofitting (six months in 2018 and three months in 2019).

Section 3.1 presents the lighting analysis of the study and Section 3.2 the same analysis for heating demands. The numerical results were based on the p-values of the ANOVA and Kruskal tests in the vectorial analysis, and on the p-values of the FANOVA in the functional analysis. Different measures are also shown to illustrate the differences between the sample groups: D_{vec}, the difference between the medians from the vectorial approach; D_{func}, the average minute difference between the mean functions; and D_{dist}, the $\mathcal{L}_2(l)$ distance between the mean functions. Additionally, to measure the functional smoothing adjustment to the raw data, the coefficient of determination R^2 is shown in the tables. The change in the variance of monitored data ($\triangle$Var) and the savings obtained with the retrofitting, calculated in relation to the initial energy demands, are also shown from both approaches. Lastly, all figures presented here were made with the R-programming software.

3.1. Lighting Analysis

The results of the electrical lighting consumption are shown in Figure 5. On the one hand, the vectorial analysis by means of box plots is shown in the first row. It can be seen that, after the refurbishment, the lighting demands of the building decreased and became more homogeneous on all floors. This is demonstrated in Table 2, where the change is quantified, and the statistical tests used (ANOVA and Kruskal) corroborate the change. In this case, the floor with the highest reduction was the third floor (2004 W less) and the floor with the lowest reduction was the second floor (489 W less). On the other hand, in the second row of Figure 5, the functional analysis, through the daily curve graph, is presented. In this case, thanks to this approach, it can be seen that, despite the decrease in the consumption, the retrofitting hardly affects the daily behaviour of the lighting consumption. As it is observable in Figure 5 that the curves have very similar shapes, with the exception on the third floor where a change in the lighting schedule was implemented. Table 2 presents the FANOVA results. The similarity of the samples is rejected on every floor. However, the impact varies among floors, as shown in Table 2. The first floor was the most benefited (2032 W less per minute on average) and the third floor the least benefited (400 W less per minute on average). The differences between the two analyses, in absolute terms, have been noticed: with the vectorial analysis the third floor obtained the highest reduction, while with functional analysis it obtained the lowest reduction. As shown in the second row of Figure 5, on the third floor, the reduction was concentrated on the last hours of the day; however, on average during the day, this reduction was lower. The vectorial approach does not take this fact into account because it distorts the sample with the calculation of daily averages.

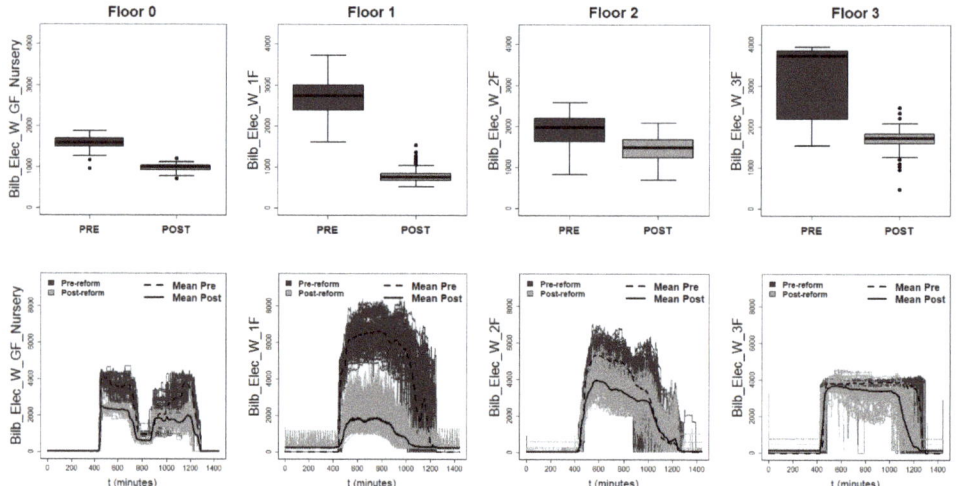

Figure 5. Analysis of the electrical lighting consumption on each floor measured in W. In the first row, the vectorial results (in form of box plots) are presented. In the second row, instead, the functional data are represented with the respective mean functions. The data are divided into winter days before and after the refurbishment.

Table 2. Numerical results on each floor for lighting consumption. The vectorial results are presented with D_v measuring the difference between medians. The functional results, accompanied with the average minute difference (D_{func}), the $\mathcal{L}_2(l)$ distance between the mean functions (D_{dist}) and the smoothing adjustment (R^2), are also presented. For both analyses, the change in the variability of the data ($\triangle$Var) and the electrical savings are displayed. Lastly, the p-values for the tests, both vectorial and functional, are displayed in this table.

	Electrical Consumption									
	Vectorial Analysis					Functional Analysis				
	P_{anova}	$P_{kruskal}$	D_{vec} (W)	$\triangle$Var	Savings	P_{fanova}	D_{func} (W)	D_{dist} ($\mathcal{L}_2(l)$)	$\triangle$Var	Savings	R^2
Floor 0	≈0	≈0	−587.48	−72.63%	36.96%	≈0	−609.69	34536.02	−71.80 %	38.11%	0.9909
Floor 1	≈0	≈0	−1987.68	−94.47%	72.68%	≈0	−2032.31	116400.04	−89.54%	73.15%	0.9903
Floor 2	≈0	≈0	−489.05	−43.49%	24.83%	≈0	−434.68	29537.58	−55.54%	22.50%	0.9929
Floor 3	≈0	≈0	−2004.80	−95.32%	53.82%	≈0	−400.82	35603.07	+17.29%	18.18%	0.9915

On the other hand, the effects of the refurbishment on the illuminance conditions were studied. Figure 6 shows the analysis for the illuminance levels. Through vectorial analysis, an impact is also detected, but is different on each floor. In general, the illuminance level was improved. Table 3 indicates that the floors with the biggest increase in illuminance were the first and third floors (356 lx and 414 lx more, respectively). Ground floor was the only one with an illuminance reduction from this point of view (270 lx less). In this case, this reduction is related to the shading that was installed to have a better protection against natural light (see Table 3). Furthermore, observing the functional illuminance curves floor by floor shown in the second row of Figure 6, the conclusion is the same: there was also an improvement in the illuminance levels, and the biggest increase of illuminance took place on the first and third floors. From this point of view, Table 3 shows that the increments were about 174 lx on the first floor and 204 lx on the third floor, on average, every minute. Once again, FDA makes it possible to see that the behaviour of the illuminance was maintained on all floors. As shown in Table 3, FDA detects that the illuminance changes on the second floor, while the vectorial approach fails in this detection (the p-values obtained are bigger than 0.05).

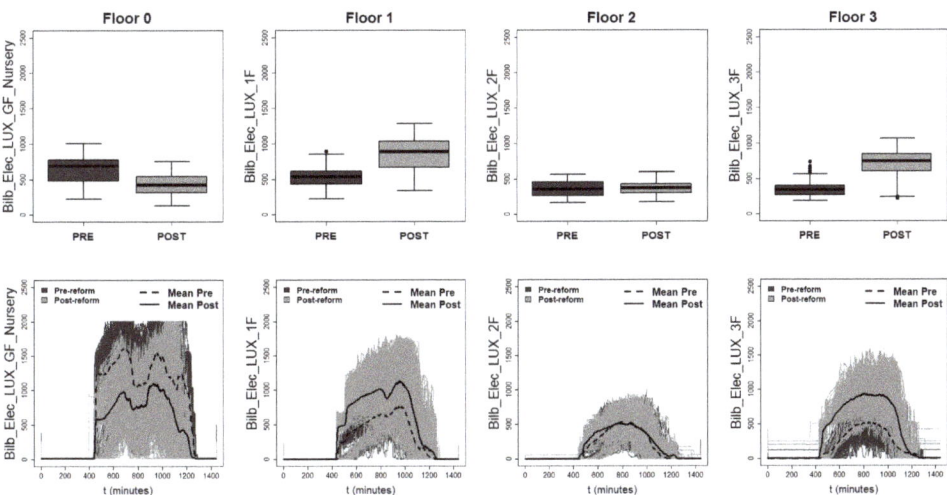

Figure 6. Analysis of the illuminance on each floor measured in lx. In the first row, the vectorial results (in form of box plots) are presented. In the second row, instead, the functional data represented with the respective mean functions. The data are divided into winter days before and after the refurbishment.

Table 3. Numerical results on each floor for illuminance. The vectorial results are presented with D_{vec} measuring the difference between medians. The functional results, accompanied with the average minute difference (D_{func}), the $\mathcal{L}_2(l)$ distance between the mean functions (D_{dist}) and the smoothing adjustment (R^2), are also presented. For both analyses, the change in the variability of the data is displayed ($\triangle$Var). Lastly, the p-values for the tests, both vectorial and functional, are displayed in this table.

			Illuminance						
	Vectorial Analysis				Functional Analysis				
	P_{anova}	$P_{kruskal}$	D_{vec} (lx)	$\triangle$Var	P_{fanova}	D_{func} (lx)	D_{dist} ($\mathcal{L}_2(l)$)	$\triangle$Var	R^2
Floor 0	≈0	≈0	−266.22	−42.27%	≈0	−286.81	15274.23	−12.51%	0.9909
Floor 1	≈0	≈0	+356.12	+62.88%	≈0	+174.78	9379.46	+59.44%	0.9913
Floor 2	0.217	0.278	+9.46	−27.78%	≈0	+23.33	1608.27	−9.29%	0.9904
Floor 3	≈0	≈0	+414.86	+81.24%	≈0	+204.32	11140.84	+66.30%	0.9914

Analysing the graphs shown in Figures 5 and 6, it can be seen that the retrofitting reduced the electrical consumption of lighting while the illuminance levels were improved or maintained in proper levels. The level of illuminance on the first and third floors was improved mainly due to the replacement by LED technology, as shown in Figure 6 and Table 3. In the case of the second floor, although it obtained an average reduction of 435 W, the illuminance level was maintained according to European illuminance regulations. This regulation states that the optimal lighting level in offices is 500 lx, a level already reached before the renovation. In the case of the ground floor, as mentioned above, it had to be analysed individually. The mean function of illuminance before retrofitting on this floor had peaks above 1500 lx (see Figure 6), which indicates a high influence of natural light. After the retrofitting, the shading that was installed to protect from sunlight achieved a reduction, on average, of 286 lx each minute, as it can be seen in Table 3 (266 lx, observing the vectorial results). Furthermore, the electrical consumption associated to lighting on the ground floor is also reduced. Table 2 shows that the vectorial reduction was 587 W, and the functional reduction 609 W, on average, every minute.

Generally, monitored lighting consumption data after retrofitting have less dispersion (see $\triangle$Var in Table 2). This means that the monitored lighting data are more homogeneous. In this case, the reduction of the data variability reaches values higher than 30% from both methods. However, the functional analysis detects an increase of 17% in the data dispersion of the third floor, as shown in Table 2. The functional graph of this floor in Figure 5 supports this result. The vectorial approach, which summarises the days with the mean, distorts the results by considering false conclusions such as, in this case, that the data variance decreased for all floors (see Table 2). On the other hand, the homogeneity of the monitored illuminance data after the renovation is different depending on the floor. Table 3 shows that the floors where the data dispersion was reduced are the ground and the second floors.

From an energetic point of view, the relative electrical consumption savings after retrofitting are also calculated. Table 2 shows that savings of more than 18% were obtained on every floor with the functional analysis and more than 24% with the vectorial analysis. The first floor is the most benefited floor with savings around 73% from the two approaches applied. However, there are differences between the results of the vectorial analysis and the functional analysis. This is shown in Figures 5 and 6 and, specifically, in Table 2 where the savings are presented. The savings calculation of functional analysis are more accurate because it is based on the areas under the curves representing the mean functions, taking into account the entire daily behaviour. The vectorial results, instead, only quantify the differences between the medians of the samples.

3.2. Heating Analysis

The results show that the refurbishment had a significant impact in the heating demands of the building. First, observing the box plots from the vectorial analysis, as shown in the first row of Figure 7, it is clear that the heating demand was reduced on all floors. This fact is supported by vectorial tests that reject similarity between samples on all floors (see Table 4). After that, with the functional graphs, as shown in the second row of Figure 7, it is also appreciated that the heating demand curves after refurbishment are below the initial curves. This is also proved by the functional tests that reject all the sample similarity hypothesis (see Table 4). The results shown in Table 4 change from vectorial to functional analysis. Both analyses come to the same conclusions, but the magnitude of change is different for each one. With vectorial approach, the reduction per floor ranges from 3667 W as the highest reduction on the third floor to 1057 W as the lowest on the first floor. In contrast, with the functional analysis, although the highest and lowest reduction were on the same floors, the values of the reduction are not the same. The heating demand, each minute, was reduced on average 3918 W on third floor and 1456 W on first floor. Therefore, the relative heating savings were significant (see Table 4). The floor most benefited was the third floor; both analysis obtained savings higher than a 30% on this floor. Again, the ground floor had to be analysed individually because, in addition to the installation of shading in 2017 that prevents solar gains, a false ceiling was already insulated on this floor in 2015. Thus, it was possible to obtain a saving of about 12% of the initial heating demands from vectorial results and about 17% from functional results (see Table 4).

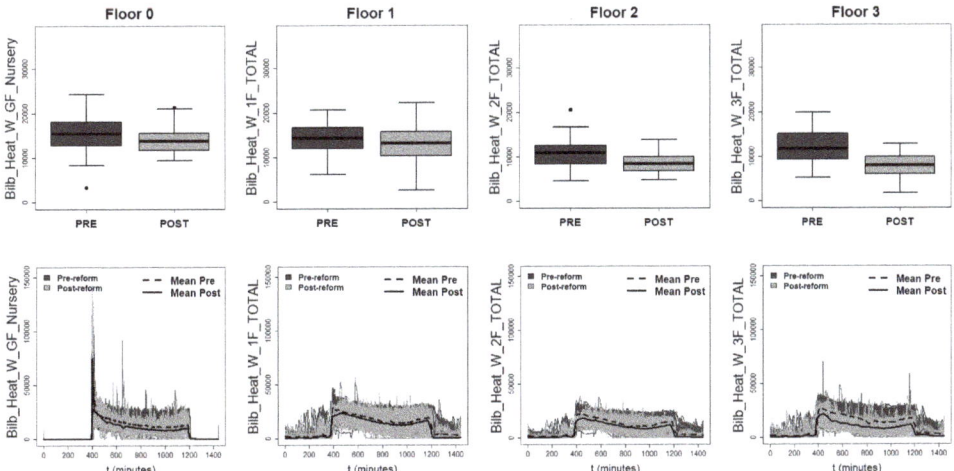

Figure 7. Analysis of the heating demands of each floor measured in W. In the first row, the vectorial results (in form of box plots) are presented. In the second row, instead, the functional data are represented with the respective mean functions. The data are divided into winter days before and after the refurbishment.

Table 4. Numerical results on each floor results for heating demands. The vectorial results are presented with D_{vec} measuring the difference between medians. The functional results, accompanied with the average minute difference (D_{func}), the $\mathcal{L}_2(l)$ distance between the mean functions (D_{dist}) and the smoothing adjustment (R^2), are also presented. For both analyses, the change in the variability of the data ($\triangle$Var) and the heating savings are displayed. Lastly, the p-values for the tests, both vectorial and functional, are calculated.

	Heating Demand									
	Vectorial Analysis					Functional Analysis				
	P_{anova}	$P_{kruskal}$	D_{vec} (W)	$\triangle$Var	Savings	P_{fanova}	D_{func} (W)	$D_{dist}(\mathcal{L}_2(l))$	$\triangle$Var	Savings	R^2
Floor 0	1.761×10^{-6}	7.047×10^{-6}	−1838.12	−35.68%	11.86%	≈0	−1455.83	158722.42	−30.66%	17.36%	0.9901
Floor 1	0.018	0.02	−1057.67	+24.71%	7.46%	≈0	−1975.15	95981.90	−23.89%	16.97%	0.9908
Floor 2	≈0	≈0	−2457.87	−43.99%	22.73%	≈0	−2158	96002.01	−43.48%	23.60%	0.9914
Floor 3	≈0	≈0	−3667.06	−60.22%	31.49%	≈0	−3917.87	185968.48	−51.68%	35.51%	0.9911

In the case of the vectorial analysis, Table 4 shows that, in general, the measurements are less variable in general (between 35% and 60% less). On the first floor, instead, this method detects an increase in data dispersion after the retrofitting. However, this fact is not supported by the functional approach (see Table 4). Figure 7 shows in its second row that the heating curves on this floor are similar or even less variable. In the case of functional analysis results, as shown in Table 4, the variation of the measurements is lower on all floors (between 23% and 51% less). The functional method is demonstrated to be more accurate and provides more information. For instance, functional analysis detects the demand peaks on ground floor in the morning when the heating starts, as it can be seen in second row of Figure 7. With the vector analysis, this information is lost as shown in first row of Figure 7.

The possible consequences of the retrofitting on the building temperatures are also studied. Both the vectorial approach and the functional approach conclude that the temperatures increased on every floor except on the third floor (see Figure 8 and Table 5). The tests do not detect any change in the average indoor temperature on this floor. This is probably because this floor is a large space with low occupancy where the temperatures were stable before and after refurbishment. This is supported, on the one hand, by the functional analysis in Figure 8 where the curves are almost overlapping. On the other hand, Table 5 shows that the temperature on the third floor increased very slightly and the FANOVA test does not detect a significant change (p-value = 0.17). On the contrary, the ground, first and second floors had higher temperatures after retrofitting, as shown in Figure 8 and Table 5. The increase, depending on the floor and method, was between 0.5 and 2 °C. The reason is that after the refurbishment the building is more insulated, heat losses are reduced and it is easier to keep it warmer. Moreover, the temperature set point have been increased in the common zones. Only with FDA it is detected that the retrofitting succeeds to reduce the influence of natural light on the ground floor temperatures. In the second row of Figure 8, it is observed that temperatures on the ground floor after refurbishment do not have peaks at the end of the day due to solar radiation. Finally, both analyses show that the homogeneity of monitored data is significantly improved. Table 5 shows that, on each floor and from both approaches, there is more homogeneity in the measurements related to the building's indoor temperatures.

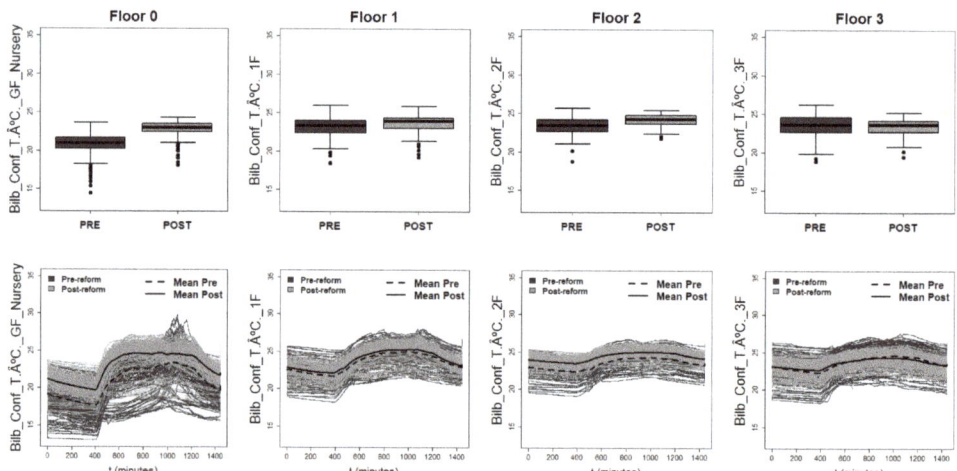

Figure 8. Analysis of the temperatures on each floor measured in °C. In the first row, the vectorial results (in form of box plots) are presented. In the second row, instead, the functional data are represented with the respective mean functions. The data are divided into winter days before and after the refurbishment.

Table 5. Numerical results on each floor indoor temperatures. The vectorial results are presented with D_{vec} measuring the difference between medians. The functional results, accompanied with the average minute difference (D_{func}), the $\mathcal{L}_2(l)$ distance between the functional mean (D_{dist}) and the smoothing adjustment (R^2), are also presented. For both analyses, the change in the variability of the data is displayed ($\triangle$Var). Lastly, the p-values for the tests, both vectorial and functional, are calculated.

	Indoor Temperatures								
	Vectorial Analysis				Functional Analysis				
	P_{anova}	$P_{kruskal}$	D_{vec} (W)	$\triangle$Var	P_{fanova}	D_{func} (W)	D_{dist} ($\mathcal{L}_2(l)$)	$\triangle$Var	R^2
Floor 0	≈ 0	≈ 0	+1.92	−62.75%	≈ 0	+1.95	75.09	−40.10%	0.9918
Floor 1	0.001	0.004	+0.55	−45.33%	≈ 0	+0.35	13.43	−37.77%	0.9970
Floor 2	≈ 0	≈ 0	+0.76	−53.31%	≈ 0	+0.92	35.29	−60.68%	0.9976
Floor 3	0.39	0.274	−0.072	−50.65%	0.17	+0.002	7.97	−63.15%	0.9959

As in the electrical analysis, the relative savings obtained in the heating analysis were calculated, as shown in Table 4. These savings reached values of more than 30% (in particular on third floor), and, in this case, with the functional approach are higher than with the vectorial approach. The floor with the smallest relative saving, observing the results of both methods, was the first floor, but with the vectorial method the reduction was almost 8% and with the functional method almost 17%. In this case, the results are more homogeneous among floors; there is not much difference from floor to floor. As expected, the form of the heating demands curves, before and after retrofitting, is the same although the values after retrofitting decreased (see Figure 7). Similar behaviour is appreciated in the temperature curves (see Figure 8).

After retrofitting the indoors temperature increased on all floors, a ventilation system with exterior air was installed and the internal gains were reduced with LED lighting. These changes should contribute to an increase of the heating demand. However, Figure 7 and Table 4 show that the heating demand on each floor has been reduced, demonstrating the effectiveness of façade insulation. Besides the decrease on the heating demands, indoor temperatures in the building are maintained or even increased, as demonstrated in Figure 8 and Table 5.

4. Conclusions

A new application of FDA and a new methodology to assess the impact of retrofitting in buildings are presented in this paper. The study was conducted by analysing monitored data of lighting consumption, heating demands, illuminance levels and indoor temperatures of the Rectorate building of the University of the Basque Country (Spain). These analyses aimed to detect an impact on the measured variables and to quantify the changes achieved with retrofitting. The methodology used in this study is based, on the one hand, in the functional analysis contrasting the distance between mean functions of monitored samples before and after retrofitting. On the other hand, as a comparison, the classical or vectorial approach was carried out measuring the dissimilarities between sample medians.

The presented method contributes with advantages over the already existing research in the topic of building retrofitting evaluation. Some research evaluated the effect of the refurbishment based on environmental indicators. Other studies give a measure of the heat losses before and after retrofitting, and other studies focus on the heating demands. One of the advantages of the proposal of using functional analysis is that it can be applied to evaluate different building variables such as temperatures, lighting levels, electrical consumption or heating demands. Consequently, it can be used to search for relationship or effects between variables. For instance, if the heating demand is not reduced as expected, the evolution of other variables can be observed to look for the cause. Moreover, a daily based analysis can be done, evaluating the peculiarities of some days in the performance of the studied variables.

The research contribution of this paper is the application of a mathematical method such as functional analysis to evaluate the impact that a building retrofitting had in its energy performance. There are few studies that present evaluations of retrofitting actions in buildings with monitored data and the employed methods are based in vector-based data approaches. The benefits of applying FDA to contrast and measure the similarity between samples of monitored data before and after a retrofitting were demonstrated. An advantage of FDA is that it is not restricted to certain characteristics of the data distribution, thus it is not necessary to test its normality. It considers complete time units, working with the time set in a continuous way, without having to summarise them, which is beneficial to evaluate monitored building variables. In addition, the outlier detection takes into account constant deviations as a reason to identify an outlier, even though it does not surpass the cut-off criterion.

Furthermore, the variables used in this study are commonly analysed in different types of buildings. The methodology presented here can be applied to assess the energy and thermal performance of different buildings, such as industrial, residential, educational or office buildings. Functional analysis can be applied, with variables or tools different from those used here, to evaluate different aspects of the studied building. Thus, this method could also be used to evaluate monitored variables in thermal facilities within an energy efficiency framework. The effectiveness and usefulness of the functional approach to evaluate variables that affect the energy efficiency was demonstrated in this study.

The results illustrate the greater accuracy of FDA in detecting if there was a significant change in the studied variables in comparison with vectorial analyses results. In the illuminance analysis of the second floor, the FANOVA detected a change (p-value $\approx$ 0) while the vectorial ANOVA did not (p-value = 0.22). Additionally, FDA could detect that most of the lighting consumption reduction on the third floor was concentrated in the last hours of the day. In addition, it has been demonstrated that FDA can provide trustable information about the dispersion of the data. In the lighting consumption analysis of the third floor, the greater dispersion of data after retrofitting was only identified with FDA. Moreover, the representation of the monitored variables in a continuous way throughout the day shows that the FDA allows a greater accuracy and a better adjustment to reality than the vectorial methods. It allows identifying average patterns of the studied variables and it has the potential for detecting anomalous behaviours of monitored variables. In the analysis of the heating demand of the ground floor, the FDA reported a peak in the early hours of the day (around 7 p.m.), both before and after the retrofitting, which the vectorial analysis did not notice.

From an energetic point of view, the conclusion is that the refurbishment carried out in the building under study had a significant impact on its energy performance. The main aim of the retrofitting was to reduce the heat losses. This goal was achieved, as demonstrated by the decrease in the heating demands of the entire building even though the temperatures were increased, a ventilation system being installed and the LED lighting reducing the internal heat gains. The analysis also shows that the illuminance levels were improved on all floors, only decreasing on the ground floor as a shading was installed to prevent from direct sunlight. According to FDA results, the heating demands were reduced 17% on the ground and first floor, 24% on the second floor and 36% on the third floor. Moreover, the reduction in lighting consumptions were 38% for the ground floor, 73% for the first floor and around 20% for the second and third floors.

Author Contributions: Mathematical methodology, M.M.C.; engineering knowledge, S.M.M. and E.G.Á.; and data collection, P.E.O. and A.E.G. All authors have read and agreed to the published version of the manuscript.

Funding: This paper was funded by the Spanish Government (Science, Innovation and Universities Ministry) under the project RTI2018-096296-B-C21.

Acknowledgments: This paper was supported by the Spanish Government (Science, Innovation and Universities Ministry) under the project RTI2018-096296-B-C21.

Conflicts of Interest: The authors declare no conflict of interest.

References

1. Foucquier, A.; Robert, S.; Suard, F.; Stéphan, L.; Jay, A. State of the art in building modelling and energy performances prediction: A review. *Renew. Sustain. Energy Rev.* **2013**, *23*, 272–288. [CrossRef]
2. Energy Information Administration. International Energy Outlook 2019. Available online: https://www.eia.gov/ieo (accessed on 21 March 2020).
3. RCP policy: public health and health inequality. Every breath we take: the lifelong impact of air pollution. Royal College of Physicians (RCPCH). 2016. Available online: https://www.rcplondon.ac.uk/projects/outputs/every-breath-we-take-lifelong-impact-air-pollution (accessed on 21 March 2020).
4. Wang, L.; Pereira, N.; Hung, Y. *Air Pollution Control Engineering*; Humana Press: New York, NY, USA, 2004.
5. EEA. *Air Quality in Europe*; European Environmental Agency: Copenhagen, Denmark, 2019.
6. Frances Bean, J.; Dorizas, V.; Bourdakis, E.; Staniaszek, D.; Pagliano, L.; Roscetti, A. *Future-Proof Buildings for all Europeans—A Guide to Implement the Energy Performance of Buildings Directive*; Buildings Performance Institute Europe (BPIE): Brussels, Belgium, 2019.
7. Cabeza, L.F.; Rincón, L.; Vilariño, V.; Castell, A. Life cycle assessment (LCA) and life cycle energy analysis (LCEA) of buildings and the building sector: A review. *Renew. Sustain. Energy Rev.* **2014**, *29*, 394–416. [CrossRef]
8. Santamouris, M.; Pavlou, C.; Doukas, P.; Mihalakakou, G.; Synnefa, A.; Hatzibiros, A.; Patargias, P. Investigating and analysing the energy and environmental performance of an experimental green roof system installed in a nursery school building in Athens, Greece. *Energy* **2007**, *32*, 1781–1788. [CrossRef]
9. Chidiac, S.E.; Catania, E.J.; Morofky, E.; Foo, S. Effectiveness of single and multiple energy retrofit measures on the energy consumption of office buildings. *Energy* **2011**, *36*, 5037–5052. [CrossRef]
10. Zmeureanu, R. Assessment of the energy savings due to the building retrofit. *Build. Environ.* **1990**, *25*, 95–103. [CrossRef]
11. Asadi, E.; Da Silva, M.G.; Antunes, C.H.; Dias, L. Multi-objective optimization for building retrofit strategies: A model and an application. *Energy Build.* **2012**, *44*, 81–87. [CrossRef]
12. Yalcintas, M. Energy-savings predictions for building-equipment retrofits. *Energy Build.* **2008**, *40*, 2111–2120. [CrossRef]
13. Tobias, L.; Vavaroutsos, G. *Retrofitting Buildings to be Green and Energy-Efficient: Optimizing Building Performance, Tenant Satisfaction, and financial Return*; Urban Land Institute: Washington, DC, USA, 2012.
14. Hamburg, A.; Kalamees, T. How well are energy performance objectives being achieved in renovated apartment buildings in Estonia? *Energy Build.* **2019**, *199*, 332–341. [CrossRef]
15. Ardente, F.; Beccali, M.; Cellura, M.; Mistretta, M. Energy and environmental benefits in public buildings as a result of retrofit actions. *Renew. Sustain. Energy Rev.* **2011**, *15*, 460–470. [CrossRef]
16. Mohammadpourkarbasi, H.; Sharples, S. Eco-retrofitting very old dwellings: current and future energy and carbon performance for two UK cities, Plea 2013. In Proceedings of the 29th Conference, Munich, Germany, 10–12 September 2013.
17. Beccali, M.; Cellura, M.; Fontana, M.; Longo, S.; Mistretta, M. Energy retrofit of a single-family house: Life cycle net energy saving and environmental benefits. *Renew. Sustain. Energy Rev.* **2013**, *27*, 283–293. [CrossRef]
18. Famuyibo, A.A.; Duffy, A.; Strachan, P. Achieving a holistic view of the life cycle performance of existing dwellings. *Build. Environ.* **2013**, *70*, 90–101. [CrossRef]
19. Ficco, G.; Iannetta, F.; Ianniello, E.; dAmbrosio Alfano, F.R.; DellIsola, M. U-value in situ measurement for energy diagnosis of existing buildings. *Energy Build.* **2015**, *104*, 108–121. [CrossRef]
20. Uriarte, I.; Erkoreka, A.; Giraldo-Soto, C.; Martin, K.; Uriarte, A.; Eguia, P. Mathematical development of an average method for estimating the reduction of the Heat Loss Coefficient of an energetically retrofitted occupied office building. *Energy Build.* **2019**, *192*, 101–122. [CrossRef]
21. Zavadskas, E.; Kaklauskas, A.; Turskis, Z.; Kalibatas, D. An approach to multi-attribute assessment of indoor environment before and after refurbishment of dwellings. *J. Environ. Eng. Landsc. Manag.* **2009**, *17*, 5–11. [CrossRef]
22. Febrero, M.; Galeano, P.; Wenceslao, G. Outlier detection in functional data by depth measures, with application to identify abnormal NOx levels. *Environmetrics* **2008**, *19*, 331–345. [CrossRef]

23. Horváth, L.; Kokoszka, P. *Inference for Functional Data with Applications*; Springer: Berlin/Heidelberg, Germany, 2012. [CrossRef]
24. Martínez, J.; Saavedra, A.; García, P.; Piñeiro, J.; Iglesias, C.; Taboada, J.; Sancho, J.; Pastor, J. Air quality parameters outliers detection using functional data analysis in the Langreo urban area (Northern Spain). *Appl. Math. Comput.* **2014**, *241*, 1–10. [CrossRef]
25. Pi neiro, J.; Torres, J.; García, P.; Alonso, J.; Mu niz, C.; Taboada, J. Analysis and detection of functional outliers in waterquality parameters from different automated monitoring stationsin the Nalón River Basin (Northern Spain). *Environ. Sci. Pollut. Res.* **2015**, *22*, 387–396. [CrossRef]
26. Martínez, J.; Pastor, J.; Sancho, J.; McNabola, A.; Martínez, M.; Gallagher, J. A functional data analysis approach for the detection of air pollution episodes and outliers: A case study in Dublin, Ireland. *Mathematics* **2020**, *8*, 225. [CrossRef]
27. Sancho, J.; Martínez, J.; Pastor, J.; Taboada, J.; Piñeiro, J.; García Nieto, P. New methodology to determine air quality in urban areas based on runs rules for functional data. *Atmos. Environ.* **2014**, *83*, 185–192. [CrossRef]
28. Sancho, J.; Iglesias, C.; Piñeiro, J.; Martínez, J.; Pastor, J.; Araújo, M.; Taboada, J. Study of water quality in a spanish river based on statistical process control and functional data analysis. *Math. Geosci.* **2016**, *48*, 163–186. [CrossRef]
29. Iglesias, C.; Sancho, J.; Piñeiro, J.I.; Martínez, J.; Pastor, J.J.; Taboada, J. Shewhart-type control charts and functional data analysis for water quality analysis based on a global indicator. *Desalin. Water Treat.* **2016**, *57*, 2669–2684. [CrossRef]
30. Dombeck, D.; Graziano, M.; Tank, D. Functional clustering of neurons in motor cortex determined by cellular resolution imaging in awake behaving mice. *J. Neurosis.* **2009**, *29*, 13751–13760. [CrossRef]
31. Martínez, J.; Ordoñez, C.; Matìas, J.M.; Taboada, J. Determining noise in an aggregates plant using functional statistics. *Hum. Ecol. Risk Assess.* **2011**, *17*, 521–533. [CrossRef] [PubMed]
32. Ordóñez, C.; Martínez, J.; Saavedra, A.; Mourelle, A. Intercomparison Exercise for Gases Emitted by a Cement Industry in Spain: A Functional Data Approach. *J. Air Waste Manag. Assoc. 1995* **2011**, *61*, 135–141. [CrossRef]
33. Sancho, J.; Pastor, J.; Martínez, J.; García, M. Evaluation of Harmonic Variability in Electrical Power Systems through Statistical Control of Quality and Functional Data Analysis. In *Procedia Engineering*; The Manufacturing Engineering Society: Southfield, MI, USA, 2013; Volume 63, pp. 295–302. [CrossRef] [PubMed]
34. Wu, D.; Huang, S.; Xin, J. Dynamic compensation for an infrared thermometer sensor using least-squares support vector regression (LSSVR) based functional link artificial neural networks (FLANN). *Meas. Sci. Technol.* **2008**, *19*, 105202.1–105202.6. [CrossRef]
35. Ordoñez, C.; Martínez, J.; Cos Juez, J.; Sánchez Lasheras, F. Comparison of GPS observations made in a forestry setting using functional data analysis. *Int. J. Comput. Math.* **2012**, *89*, 402–408. [CrossRef]
36. Müller, H.G.; Sen, R.; Stadtmüller, U. Functional data analysis for volatility. *J. Econometr.* **2011**, *165*, 233–245. [CrossRef]
37. López, M.; Martínez, J.; Matías, J.M.; Taboada, J.; Vilán, J.A. Functional classification of ornamental stone using machine learning techniques. *J. Comput. Appl. Math.* **2010**, *234*, 1338–1345. [CrossRef]
38. López, M.; Martínez, J.; Matías, J.M.; Taboada, J.; Vilán, J.A. Shape functional optimization with restrictions boosted with machine learning techniques. *J. Comput. Appl. Math.* **2010**, *234*, 2609–2615. [CrossRef]
39. Flores, M.; Naya, S.; Fernández-Casal, R.; Zaragoza, S.; Raña, P.; Tarrío-Saavedra, J. Constructing a Control Chart Using Functional Data. *Mathematics* **2020**, *8*, 58. [CrossRef]
40. Warmenhoven, J.; Harrison, A.; Robinson, M.A.; Vanrenterghem, J.; Bargary, N.; Smith, R.; Cobley, S.; Draper, C.; Donnelly, C.; Pataky, T. A force profile analysis comparison between functional data analysis, statistical parametric mapping and statistical non-parametric mapping in on-water single sculling. *J. Sci. Med. Sport* **2018**, *21*, 1100–1105. [CrossRef]
41. Ordóñez, C.; Martínez, J.; Matías, J.M.; Reyes, A.N.; Rodríguez-Pérez, J.R. Functional statistical techniques applied to vine leaf water content determination. *Math. Comput. Model.* **2010**, *52*, 1116–1122. [CrossRef] [PubMed]

42. Eisenhart, C. The Assumptions Underlying the Analysis of Variance. *Int. Biometr. Soc.* **1947**, *3*, 1–21. [CrossRef]
43. Montgomery, D. *Design and Analysis of Experiments*, 8th ed.; John Wiley & Sons, Inc.: Hoboken, NJ, USA, 2013. [CrossRef]
44. Kotz, S.; Johnson, N.L. *Breakthroughs in Statistics. Volume II. Methodology and Distribution*; Springer: Berlin/Heidelberg, Germany, 1993; Volume 2.
45. Vikneswaran. *An R companion to "Experimental Design"*; Vikneswaran: Duxbury, MA, USA, 2005.
46. Crawley, M. *The R Book*, 2nd ed.; John Wiley & Sons: Hoboken, NJ, USA, 2013.
47. Theodorsson-Norheim, E. Kruskal-Wallis test: BASIC computer program to perform nonparametric one-way analysis of variance and multiple comparisons on ranks of several independent samples. *Comput. Meth. Prog. Biomed.* **1986**, *23*, 57–62.
48. Ostertagová, E.; Ostertag, O.; Kovác, J. Methodology and application of the Kruskal-Wallis test. *Mech. Mater.* **2014**, *611*, 115–120. [CrossRef]
49. Van Hecke, T. Power study of anova versus Kruskal Wallis test. *Stat. Manag. Syst.* **2012**, *15*, 241–247. [CrossRef]
50. Cuevas, A.; Febrero, M.; Fraiman, R. An anova test for functional data. *Comput. Stat. Data Anal.* **2004**, *47*, 111–122.
51. Tarrío-Saavedra, J.; Naya, S.; Francisco-Fernández, M.; Artiaga, R.; Lopez-Beceiro, J. Application of functional ANOVA to the study of thermal stability of micro–nano silica epoxy composites. *Chemometr. Intell. Lab. Syst.* **2011**, *105*, 114–124. [CrossRef]
52. Kaufman, C.G.; Sain, S.R. Bayesian functional (ANOVA) modeling using Gaussian process prior distributions. *Bayes. Anal.* **2010**, *5*, 123–149. [CrossRef]
53. Wang, J.L.; Chiou, J.M.; Müller, H.G. Functional Data Analysis. *Ann. Rev. Stat. Appl.* **2016**, *3*, 257–295. [CrossRef]
54. Kramosil, I.; Michálek, J. Fuzzy metrics and statistical metric spaces. *Kybernetika* **1975**, *11*, 336–344. [CrossRef]
55. Ramsay, J.; Silverman, B. *Applied Functional Data Analysis: Methods and Cae Studies*; Springer: Berlin/Heidelberg, Germany, 2002.
56. Walz, M.; Zebrowski, T.; Küchenmeister, J.; Busch, K. B-spline modal method: A polynomial approach compared to the Fourier modal method. *Opt. Express* **2013**, *21*, 14683–14697. [CrossRef]
57. Muñiz, C.; García, P.; Alonso, J.; Torres, J.; Taboada, J. Detection of outliers in water quality monitoring samples using functional data analysis in San Esteban estuary (Northern Spain). *Sci. Total Environ.* **2012**, *439*, 54–61. [CrossRef] [PubMed]
58. Martínez, J.; García, P.; Alejano, L.; Reyes, A. Detection of outliers in gas emissions from urban areas using functional data analysis. *J. Hazard. Mater.* **2011**, *186*, 144–149. [CrossRef] [PubMed]
59. Fraiman, R.; Muniz, G. Trimmed means for functional data. *TEST Off. J. Spanish Soc. Stat. Operat. Res.* **2001**, *10*, 419–440. [CrossRef]
60. Cuevas, A.; Febrero, M.; Fraiman, R. Robust estimationand classification for functional data via projection-based notions. *Comput. Stat.* **2007**, *22*, 481–496. [CrossRef]
61. Cuevas, A.; Febrero, M.; Fraiman, R. On the use of bootstrap for estimating functions with functional data. *Comput. Stat. Data Anal.* **2006**, *51*, 1063–1074. [CrossRef]
62. Léger, C.; Romano, J. Boostrap adaptive estimation. The trimmed-mean example. *Can. J. Stat.* **1990**, *18*, 297–314. [CrossRef]
63. Hall, P.; Maesono, Y. A Weighted Bootstrap Approach to Bootstrap Iteration. *J. R. Stat. Soc. Ser. B Stat. Methodol.* **2000**, *62*, 137–144. [CrossRef]
64. Millán-Roures, L.; Epifanio, I.; Martínez, V. Detection of Anomalies in Water Networks by Functional Data Analysis. *Math. Problems Eng.* **2018**, *2018*, 13. [CrossRef]
65. Dette, H.; Derbort, S. Analysis of Variance un Non Parametric regression Models. *J. Multivar. Anal.* **2001**, *76*, 110–137. [CrossRef]
66. Maldonado, Y.; Staniswalis, J.; Irwin, L.; Byers, D. A similarity analysis of curves. *Can. J. Stat.* **2002**, *30*, 373–381. [CrossRef]

67. Cuesta-Albertos, J.A.; Febrero, M. A simple multiway ANOVA for functional data. *TEST* **2010**, *19*, 537–557. [CrossRef]
68. Zhang, J.T. Analysis of Variance for Functional Data. In *A Chapman & Hall Book*; Press, C., Ed.; Taylor & Francis Group: Abingdon, UK, 2013; Chapter 5, p. 412. [CrossRef]
69. Erkoreka, A.; García, K.; Teres-Zubiaga, J.; Del Portillo, L. In-use office building energy characterization through basic monitoring and modelling. *Energy Build.* **2016**, *119*, 256–266. [CrossRef]

© 2020 by the authors. Licensee MDPI, Basel, Switzerland. This article is an open access article distributed under the terms and conditions of the Creative Commons Attribution (CC BY) license (http://creativecommons.org/licenses/by/4.0/).

Article

Classical Lagrange Interpolation Based on General Nodal Systems at Perturbed Roots of Unity

Elías Berriochoa [1,*,†], Alicia Cachafeiro [1,*,†], Alberto Castejón [1,†] and José Manuel García-Amor [2,†]

1. Departamento de Matemática Aplicada I, Universidad de Vigo, 36201 Vigo, Pontevedra, Spain; acaste@uvigo.es
2. Departamento de Matemáticas, Instituto E. S. Valle Inclán, 36001 Pontevedra, Spain; garciaamor@edu.xunta.es
* Correspondence: esnaola@uvigo.es (E.B.); acachafe@uvigo.es (A.C.); Tel.: +34-988-387216 (E.B.); +34-986-812138 (A.C.)
† These authors contributed equally to this work.

Received: 28 February 2020; Accepted: 29 March 2020; Published: 2 April 2020

Abstract: The aim of this paper is to study the Lagrange interpolation on the unit circle taking only into account the separation properties of the nodal points. The novelty of this paper is that we do not consider nodal systems connected with orthogonal or paraorthogonal polynomials, which is an interesting approach because in practical applications this connection may not exist. A detailed study of the properties satisfied by the nodal system and the corresponding nodal polynomial is presented. We obtain the relevant results of the convergence related to the process for continuous smooth functions as well as the rate of convergence. Analogous results for interpolation on the bounded interval are deduced and finally some numerical examples are presented.

Keywords: lagrange interpolation; unit circle; nodal systems; separation properties; perturbed roots of the unity; convergence

1. Introduction

The polynomial interpolation is a classical subject that has been widely studied under different approaches like Lagrange, Hermite, Birkhoff, Pál-type interpolation and some others. Although it is obvious that the subject is important by itself, its numerous numerical applications like numerical integration or numerical derivation are not less important and indeed the polynomial interpolation continues being a subject of current research.

Lagrange interpolation is a very good tool although it is known that for this interpolation scheme and for good nodal systems such as the classical Chebyshev nodes there exists a continuous function on $[-1,1]$ for which the Lagrange interpolation polynomial diverges (see [1]). A similar problem has been posed for arbitrary arrays and it was proved in [2] that for each nodal array in $[-1,1]$, there exists a continuous function such that the Lagrange polynomial interpolation diverges almost everywhere. In any case, recalling the words written by Trefethen in his paper [3] we can say that there is nothing wrong with Lagrange polynomial interpolation. "Yet the truth is, polynomial interpolants in Chebyshev points always converge if f is a little bit smooth". As a consequence, to obtain better results one needs to assume better properties on the function to be interpolated like bounded variation or a condition on its modulus of continuity. Thus one of the most important questions in relation with the interpolation of functions is the choice of the interpolation arrays or nodal systems for which one can expect to obtain pointwise or uniform convergence to the function to be interpolated and another important issue is to determine the class of functions for which some type of convergence can be guaranteed. The nodal systems strongly normal or normal, introduced by Fejér, play an important role in the interpolation theory, although from a practical point of view, the difficulty of testing the definition makes the applications of these systems quite limited. For these systems Grünwald studies in [4] the convergence of the polynomials of Lagrange interpolation for functions satisfying a Lipschitz condition.

Most of the research obtain results on the convergence from the distribution properties of the nodal points. Indeed it was Fejér, who was the first to invert the problem, trying to deduce separation properties of the nodal systems from the interpolation results. The importance of this idea is avaled by the fact that the required interpolation properties are easily verified.

In [5] it is proved that strongly normal distributions on $[-1,1]$ give quasi uniformly nodal systems on the unit circle, that is the length of the arcs between two consecutive nodes has the order of $\frac{1}{n}$. Although the situation more widely studied corresponds to the bounded interval, there are important results in some other situations in the complex plane like the unit circumference, (see [6]). By taking into account that continuous functions on the unit circle can be approximated by Laurent polynomials, the interpolation polynomials on the unit circumference are constructed in this Laurent space. In this field of research, [7] deserves to be highlighted. There, the nodal points are the roots of complex numbers with modulus 1 and in this case it is obtained a result about convergence of the interpolants for continuous functions with a suitable modulus of continuity. Moreover, in the same paper the problem of extending the results to general nodal systems. Indeed, since the roots of complex numbers with modulus 1 can be interpreted like the zeros of the para-orthogonal polynomials with respect to the Lebesgue measure, now the natural extension is to consider the zeros of the para-orthogonal polynomials with respect to other measures.

In [8] we have generalized the results given in [7] for these new nodal systems. First we work with nodal systems characterized by fulfilling some properties of boundedness, which are suggested by those fulfilled by the roots of unimodular complex numbers, obtaining a result of convergence for continuous functions with a suitable modulus of continuity. Next, by taking into account that the zeros of the para-orthogonal polynomials with respect to measures in the Szegő class (see [9]) with analytic extension up to $|z| > 1$ satisfy the properties that we need, we obtain a similar result about convergence for these type of nodal systems.

In [10] we have studied the Lagrange interpolation process for piecewise continuous functions with suitable properties and by using as nodal points the zeros of the para-orthogonal polynomials with respect to analytic weights, which constitutes a novel approach to the Lagrange interpolation theory.

Another extension to more general nodal systems is given in [11] where it has opened a new trend to interpolation at perturbed roots of unity and the functions to be interpolated belong to the disc algebra.

Now, in the present paper we assume a distribution for the nodes that can be obtained through a perturbation of the uniform distribution and, in particular of the roots of the unity, and which is more general than that given in [11]. Thus in the present paper we start from a different point of view because we base it on properties satisfied by the nodal systems and we do not need to consider orthogonality nor para-orthogonality with respect to any measure on the unit circle. The interpolation arrays are described by a separation property and the main goal is to obtain the properties that play a role in the Lagrange process, as well as to present some relevant examples.

The organization of the paper is the following. In Section 2 we introduce the nodal systems that we use throughout all the paper and we prove the main properties that they satisfy in several propositions. Section 3 is devoted to the Lagrange interpolation problem. First we obtain the Lebesgue constant of the process and then we study the convergence of the Lagrange interpolation polynomials related to continuous functions with appropriate modulus of continuity. Secondly we analyze the rate of convergence when we deal with smooth functions, (see [12]) and we also deduce analogous behavior for interpolation on the bounded interval. The last section is devoted to give some numerical examples.

2. Some General Nodal Systems on the Unit Circle

The aim of this paper is to study interpolation problems on the unit circle $\mathbb{T} = \{z \in \mathbb{C} : |z| = 1\}$ by using nodal systems satisfying some suitable properties.

We denote the nodal polynomials by $W_n(z)$ and their zeros by $\{\alpha_{j,n}\}_{j=1}^n$, that is, we assume that $W_n(z) = \prod_{j=1}^n (z - \alpha_{j,n})$, where $|\alpha_{j,n}| = 1$ for $j = 1, \cdots, n$, and $\alpha_{j,n} \neq \alpha_{k,n}$ for $j \neq k$. For simplicity, in the sequel we omit the subscript n and we write α_j instead of $\alpha_{j,n}$ for $j = 1, \cdots, n$. We denote the length of the shortest arc between any two points of the unit circle, z_1 and z_2, by $\widehat{z_1 - z_2}$, and we use the Landau's notation for complex sequences, denoting by $a_n = \mathcal{O}(b_n)$ if $|\frac{a_n}{b_n}|$ is bounded.

Throughout all the paper we assume that the zeros $\{\alpha_j\}_{j=1}^n$ of the nodal polynomials $W_n(z)$ satisfy the following separation property: there exists a positive constant A such that for $n > \frac{A}{\pi}$ the length of the shortest arc between two consecutive nodes α_j and α_{j+1}, satisfies:

$$\widehat{\alpha_j - \alpha_{j+1}} = \frac{2\pi}{n} + \frac{A(j)}{n^2} \text{ with } |A(j)| \leq A \ \forall j = 1, \cdots, n, \tag{1}$$

where $\alpha_{n+1} = \alpha_1$, that is, $\widehat{\alpha_j - \alpha_{j+1}} = \frac{2\pi}{n} + \mathcal{O}(\frac{1}{n^2})$.

We use the same $\mathcal{O}$ to denote different sequences. Unless we mention otherwise, the bounds we obtain from (1) will be uniform.

We also consider other nodal polynomials, $\widetilde{W}_n(z)$, well connected with $W_n(z)$. If we denote $\widetilde{W}_n(z) = z^n - \lambda$, with $\lambda = \alpha_1^n$, then $\widetilde{W}_n(z) = \prod_{j=1}^n (z - \beta_j)$, where

$$\beta_j = \sqrt[n]{\lambda}, \ j = 1, \cdots, n, \text{ and } \alpha_1 = \beta_1.$$

Hence it is clear that the separation property satisfied by the zeros $\{\beta_j\}$ of $\widetilde{W}_n(z)$ is

$$\widehat{\beta_j - \beta_{j+1}} = \frac{2\pi}{n} \ \forall j = 1, \cdots, n. \tag{2}$$

In this section we obtain in several propositions the main properties of the nodal polynomials $W_n(z)$. First we recall the following well known relations between arcs and strings that we are going to use throughout the whole paper and which is based on the convex character of the arcsin function: If z_1 and z_2 belong to $\mathbb{T}$ then

$$\frac{2}{\pi}(\widehat{z_1 - z_2}) \leq |z_1 - z_2| \leq (\widehat{z_1 - z_2}). \tag{3}$$

Proposition 1. *If $\{\alpha_j\}_{j=1}^n$ and $\{\beta_j\}_{j=1}^n$, with $\alpha_1 = \beta_1$, are the nodal points satisfying the separation properties (1) and (2) and we assume they are numbered in the clockwise sense, then*

(i)
$$(j-1)(\frac{2\pi}{n} - \frac{A}{n^2}) \leq \widehat{\alpha_1 - \alpha_j} \leq (j-1)(\frac{2\pi}{n} + \frac{A}{n^2}), \text{ for } j \geq 1.$$

(ii)
$$(j+1)(\frac{2\pi}{n} - \frac{A}{n^2}) \leq \widehat{\alpha_{n-j} - \alpha_1} \leq (j+1)(\frac{2\pi}{n} + \frac{A}{n^2}), \text{ for } j \geq 0.$$

(iii)
$$\widehat{\alpha_j - \beta_j} \leq (j-1)\frac{A}{n^2}, \text{ for } j \geq 1.$$

(iv)
$$\widehat{\alpha_{n-j} - \beta_{n-j}} \leq (j+1)\frac{A}{n^2}, \text{ for } j \geq 0.$$

Proof. (i) By applying (1) we have $\frac{2\pi}{n} - \frac{A}{n^2} \leq \widehat{\alpha_1 - \alpha_2} \leq \frac{2\pi}{n} + \frac{A}{n^2}$ and for $j \geq 3$ it holds $\frac{2\pi}{n} - \frac{A}{n^2} \leq \widehat{\alpha_{j-1} - \alpha_j} \leq \frac{2\pi}{n} + \frac{A}{n^2}$. Then if we sum, it is straightforward (i).

(ii) Proceeding in the same way we get $\frac{2\pi}{n} - \frac{A}{n^2} \leq \widehat{\alpha_n - \alpha_1} \leq \frac{2\pi}{n} + \frac{A}{n^2}$ and for $j \geq 0$ it holds $\frac{2\pi}{n} - \frac{A}{n^2} \leq \widehat{\alpha_{n-j-1} - \alpha_{n-j}} \leq \frac{2\pi}{n} + \frac{A}{n^2}$. Hence, by computing the sums we have (ii).

(iii) We know that $\alpha_1 = \beta_1$ and we distinguish two possibilities depending on the position of β_j related to α_j. If

$$\widehat{\alpha_1 - \beta_j} + \widehat{\beta_j - \alpha_j} = \widehat{\alpha_1 - \alpha_j},$$

that is,

$$\widehat{\beta_1 - \beta_j} + \widehat{\beta_j - \alpha_j} = \widehat{\alpha_1 - \alpha_j}$$

and we use that $\widehat{\beta_1 - \beta_j} = (j-1)\frac{2\pi}{n}$, which is a consequence of (2), and we also take into account (i) we get

$$\widehat{\alpha_j - \beta_j} \leq (j-1)\frac{A}{n^2}.$$

The second case corresponds to $\widehat{\alpha_1 - \alpha_j} + \widehat{\alpha_j - \beta_j} = \widehat{\beta_1 - \beta_j}$ and it can be deduced in the same way.

(iv) We proceed like in (iii) distinguishing the following cases $\widehat{\alpha_{n-j} - \alpha_1} = \widehat{\alpha_{n-j} - \beta_{n-j}} + \widehat{\beta_{n-j} - \alpha_1}$ or $\widehat{\beta_{n-j} - \alpha_{n-j}} + \widehat{\alpha_{n-j} - \alpha_1} = \widehat{\beta_{n-j} - \beta_1}$. □

Notice that we can write (iii) and (iv) as follows $\widehat{\alpha_j - \beta_j} = (j-1)\mathcal{O}(\frac{1}{n^2})$ and $\widehat{\alpha_{n-j} - \beta_{n-j}} = (j+1)\mathcal{O}(\frac{1}{n^2})$.

Proposition 2. *Let us assume that the zeros of the nodal polynomials $W_n(z)$ satisfy the separation property (1). Then it holds*

$$|W_n(z)| < 2e^A, \ \forall z \in \mathbb{T}. \tag{4}$$

Moreover, it also holds

$$\frac{|W_n'(z)|}{n} < 2e^A \text{ and } \frac{|W_n''(z)|}{n^2} < 2e^A, \ \forall z \in \mathbb{T}.$$

Proof. Since $W_n(\alpha_j) = 0$ for every j, let us take $z \in \mathbb{T}$, such that z is not a nodal point. In order to obtain the result we renumber the nodes in the clockwise sense in such a way that α_1 is the nodal point nearest to z. We distinguish two cases depending on whether the node closest to z is turning in the clockwise sense or in the counter clockwise sense from z. If we assume that the situation is given in Figure 1, that is, α_1 is turning in the counter clockwise sense from z, then we have

$$\widehat{z - \alpha_1} < \frac{\widehat{\alpha_1 - \alpha_2}}{2} \leq \frac{\pi}{n} + \frac{A}{2n^2}.$$

Now we consider the polynomial $\widetilde{W}_n(z) = \prod_{j=1}^{n}(z - \beta_j)$, with $\beta_1 = \alpha_1$ and satisfying (2), introduced at the beginning of the section. Using property (1) we have $\frac{\pi}{n} + \frac{A}{2n^2} < \frac{2\pi}{n} = \widehat{\alpha_1 - \beta_2}$ and then $z - \beta_2 \neq 0$.

Now, for every j it holds $\dfrac{z-\alpha_j}{z-\beta_j}=1+\dfrac{\beta_j-\alpha_j}{z-\beta_j}$ and therefore, by using Proposition 1,

$$\left|\dfrac{z-\alpha_j}{z-\beta_j}\right|\leq 1+\dfrac{\pi}{2}\dfrac{\widehat{\beta_j-\alpha_j}}{\widehat{z-\beta_j}}\leq 1+\dfrac{\pi}{2}\dfrac{(j-1)\dfrac{A}{n^2}}{\widehat{z-\beta_j}}.$$

Besides, from property (1) we have $\widehat{z-\beta_2}>\dfrac{\pi}{2n}$ and for $j\geq 3$, $\widehat{z-\beta_j}>\dfrac{\pi}{2n}+\dfrac{(j-2)2\pi}{n}>\dfrac{(j-2)2\pi}{n}$.
Hence $\left|\dfrac{z-\alpha_j}{z-\beta_j}\right|\leq 1+\dfrac{A}{n}$.

Proceeding in the same way and taking into account that for $j\geq 0$ it holds that $\widehat{\beta_{n-j}-\alpha_{n-j}}\leq (j+1)\dfrac{A}{n^2}$ and $\widehat{z-\beta_{n-j}}>(j+1)\dfrac{2\pi}{n}$ we obtain

$$\left|\dfrac{z-\alpha_{n-j}}{z-\beta_{n-j}}\right|\leq 1+\dfrac{\pi}{2}\dfrac{\widehat{\beta_{n-j}-\alpha_{n-j}}}{\widehat{z-\beta_{n-j}}}\leq 1+\dfrac{\pi}{2}\dfrac{(j+1)\dfrac{A}{n^2}}{(j+1)\dfrac{2\pi}{n}}=1+\dfrac{A}{n}.$$

Therefore we have that

$$\dfrac{|W_n(z)|}{|\widetilde{W}_n(z)|}=\prod_{j=2}^{n}\left|\dfrac{z-\alpha_j}{z-\beta_j}\right|<e^A\ \forall z,$$

and since $|\widetilde{W}_n(z)|\leq 2$, then we get $|W_n(z)|<2e^A$.

Notice that if the node closest to z, α_1, is in the clockwise sense from z, we can proceed in a similar way. Indeed $\widehat{z-\alpha_1}<\dfrac{\widehat{\alpha_1-\alpha_n}}{2}\leq\dfrac{\pi}{n}+\dfrac{A}{2n^2}$ and since $\dfrac{\pi}{n}+\dfrac{A}{2n^2}<\widehat{\beta_n-\alpha_1}=\dfrac{2\pi}{n}$ then $z-\beta_n\neq 0$.

The second statement, related to the first and second derivatives of the nodal polynomial, is a consequence of Bernstein's theorem, (see [13]). □

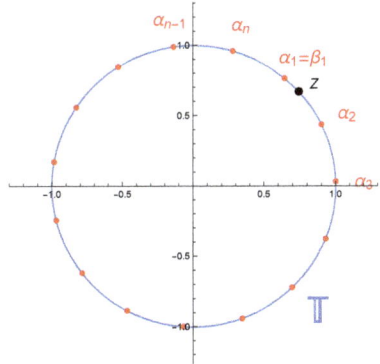

Figure 1. An arbitrary point z and the nodal system.

Proposition 3. *Let us assume that the zeros of the nodal polynomials $W_n(z)$ satisfy the separation property (1). Then there exists a positive constant $C>0$ such that for n large enough and for every $j=1,\cdots,n$, it holds that*

$$\dfrac{|W'_n(\alpha_j)|}{n}>C. \qquad (5)$$

Proof. For simplicity we take $j = 1$ and we try to bound from below $\frac{|W'_n(\alpha_1)|}{n}$.

Thus we consider the polynomial $\widetilde{W}_n(z)$ satisfying (2), that is, $\widetilde{W}_n(z) = z^n - \alpha_1^n = \prod_{j=1}^{n}(z - \beta_j)$ with $\alpha_1 = \beta_1$.

Since $\widetilde{W}'_n(z) = nz^{n-1}$ then $|\widetilde{W}'_n(\alpha_1)| = n$ and so our aim is to prove that $\frac{|W'_n(\alpha_1)|}{|\widetilde{W}'_n(\alpha_1)|} > C$.

Now, by taking into account that

$$\frac{|W'_n(\alpha_1)|}{|\widetilde{W}'_n(\alpha_1)|} = \prod_{j=2}^{n}\left|\frac{\alpha_1 - \alpha_j}{\alpha_1 - \beta_j}\right|,$$

we study the quotients $\left|\frac{\alpha_1 - \alpha_j}{\alpha_1 - \beta_j}\right|$ and $\left|\frac{\alpha_1 - \alpha_{n-j}}{\alpha_1 - \beta_{n-j}}\right|$.

On the one hand,

$$\left|\frac{\alpha_1 - \alpha_j}{\alpha_1 - \beta_j}\right| = \left|1 + \frac{\beta_j - \alpha_j}{\alpha_1 - \beta_j}\right| \geq 1 - \left|\frac{\beta_j - \alpha_j}{\alpha_1 - \beta_j}\right|,$$

and by applying (3) and Proposition 1, we have for $j \geq 2$,

$$\left|\frac{\beta_j - \alpha_j}{\alpha_1 - \beta_j}\right| \leq \frac{\pi}{2}\frac{\widehat{\beta_j - \alpha_j}}{\widehat{\alpha_1 - \beta_j}} \leq \frac{\frac{\pi}{2}(j-1)\frac{A}{n^2}}{(j-1)\frac{2\pi}{n}} = \frac{A}{4n},$$

and therefore $\left|\frac{\alpha_1 - \alpha_j}{\alpha_1 - \beta_j}\right| \geq 1 - \frac{A}{4n}$.

On the other hand,

$$\left|\frac{\alpha_1 - \alpha_{n-j}}{\alpha_1 - \beta_{n-j}}\right| = \left|1 + \frac{\beta_{n-j} - \alpha_{n-j}}{\alpha_1 - \beta_{n-j}}\right| \geq 1 - \left|\frac{\beta_{n-j} - \alpha_{n-j}}{\alpha_1 - \beta_{n-j}}\right|$$

and since for $j \geq 0$ we have

$$\left|\frac{\beta_{n-j} - \alpha_{n-j}}{\alpha_1 - \beta_{n-j}}\right| \leq \frac{\pi}{2}\frac{\widehat{\beta_{n-j} - \alpha_{n-j}}}{\widehat{\alpha_1 - \beta_{n-j}}} \leq \frac{\frac{\pi}{2}(j+1)\frac{A}{n^2}}{(j+1)\frac{2\pi}{n}} = \frac{A}{4n},$$

then $\left|\frac{\alpha_1 - \alpha_{n-j}}{\alpha_1 - \beta_{n-j}}\right| \geq 1 - \frac{A}{4n}$. Hence

$$\frac{|W'_n(\alpha_1)|}{|\widetilde{W}'_n(\alpha_1)|} = \prod_{j=2}^{n}\left|\frac{\alpha_1 - \alpha_j}{\alpha_1 - \beta_j}\right| \geq (1 - \frac{A}{4n})^{n-1},$$

that is, $|W'_n(\alpha_1)| \geq (1 - \frac{A}{4n})^{n-1}n$. Thus, given $\varepsilon > 0$ if $C = e^{-\frac{A}{4}} - \varepsilon > 0$, then for n large enough it holds that $|W'_n(\alpha_1)| > Cn$.

Notice that for another value of j there is no any significant difference. Indeed to obtain that $|W'_n(\alpha_i)| > Cn$, we take the auxiliary polynomial $\widetilde{W}_n(z) = z^n - \alpha_i^n = \prod_{j=1}^{n}(z - \beta_j)$ with $\alpha_i = \beta_i$, we renumber the nodes as in the previous proof and we proceed in a similar way. □

Proposition 4. *Let us assume that the zeros of the nodal polynomials $W_n(z)$ satisfy the separation property (1). Then there exists a positive constant $D > 0$ such that*

$$\frac{|W_n(z)|^2}{n^2} \sum_{j=1}^{n} \frac{1}{|z-\alpha_j|^2} < D, \quad \forall z \in \mathbb{T}. \tag{6}$$

Proof. Following the same steps of the proof of Lemma 2 in [14] we have

$$|W_n(z)|^2 \sum_{j=1}^{n} \frac{1}{|z-\alpha_j|^2} = |zW_n(z)W_n'(z) + z^2(W_n''(z)W_n(z) - (W_n'(z))^2)|$$

and therefore by using (4) and its consequences in Proposition 2 we get

$$\frac{|W_n(z)|^2}{n^2} \sum_{j=1}^{n} \frac{1}{|z-\alpha_j|^2} \leq \frac{|W_n(z)|}{n} \frac{|W_n'(z)|}{n} + \frac{|W_n''(z)|}{n^2}|W_n(z)| + \frac{|W_n'(z)|^2}{n^2} <$$

$$\frac{B}{n}B + B^2 + B^2, \text{ where } B = 2e^A.$$

□

Remark 1. *The nodal systems considered in [15] satisfy condition (1). Indeed they are the para-orthogonal polynomials related to measures in the Baxter class, (see [16]). In that work it is also assumed the additional condition that the sequence $\{(\phi_n^*)'\}$ is uniformly bounded on $\mathbb{T}$, where $\{\phi_n\}$ is the sequence of monic orthogonal polynomials related to the measure and $\{\phi_n^*\}$ is the sequence of the reciprocal polynomials, (see [9]). In that situation studied in [15], properties (4)–(6) also hold. Now, in the present paper we start from a different point of view because we base it on properties satisfied by the nodal systems and we do not need to consider orthogonality nor para-orthogonality with respect to any measure.*

3. Lagrange Interpolation. Convergence in Case of Smooth Continuous Functions

To compute the interpolation polynomials, first we recall some well known definitions related to interpolation problems on the unit circle. We work in the space of Laurent polynomials and, in particular, in the subspaces $\Lambda_{p,q}[z] = span\{z^k : p \leq k \leq q\}$, with p and q integers $p \leq q$.

Let us continue denoting by $\{\alpha_j\}_{j=1}^n$ the zeros of the the nodal polynomial $W_n(z)$. If $\{u_j\}_{j=1}^n$ are arbitrary complex numbers, the Laurent polynomial of Lagrange interpolation $\mathcal{L}_{-E[\frac{n}{2}],E[\frac{n-1}{2}]}(z) \in \Lambda_{-E[\frac{n}{2}],E[\frac{n-1}{2}]}[z]$ characterized by satisfying

$$\mathcal{L}_{-E[\frac{n}{2}],E[\frac{n-1}{2}]}(\alpha_j) = u_j, \text{ for } j = 1, \cdots, n,$$

has the following expression

$$\mathcal{L}_{-E[\frac{n}{2}],E[\frac{n-1}{2}]}(z) = \frac{W_n(z)}{z^{E[\frac{n}{2}]}} \sum_{j=1}^{n} \frac{\alpha_j^{E[\frac{n}{2}]}}{W_n'(\alpha_j)(z-\alpha_j)} u_j.$$

If F is a function and $u_j = F(\alpha_j)$ we denote the corresponding Laurent polynomial $\mathcal{L}_{-E[\frac{n}{2}],E[\frac{n-1}{2}]}(F,z)$. If n is odd, since $E[\frac{n-1}{2}] = E[\frac{n}{2}]$, then the interpolation polynomial $\mathcal{L}_{-E[\frac{n}{2}],E[\frac{n}{2}]}(z) \in \Lambda_{-E[\frac{n}{2}],E[\frac{n}{2}]}[z]$ and when n is even, taking into account that $E[\frac{n-1}{2}] = E[\frac{n}{2}]-1$, then the Laurent polynomial of Lagrange interpolation $\mathcal{L}_{-E[\frac{n}{2}],E[\frac{n}{2}]-1}(z) \in \Lambda_{-E[\frac{n}{2}],E[\frac{n}{2}]-1}[z]$.

Without loss of generality, to fix ideas and to simplify the notation we assume that the number of nodes is even, $2n$, in which case the interpolation polynomial $\mathcal{L}_{n,n-1}$ belongs to the space $\Lambda_{-n,n-1}$ and it can be written in terms of the fundamental polynomials as follows:

$$\mathcal{L}_{-n,n-1}(z) = \frac{W_{2n}(z)}{z^n} \sum_{j=1}^{2n} \frac{\alpha_j^n}{W'_{2n}(\alpha_j)(z-\alpha_j)} u_j. \tag{7}$$

In order to compute the interpolation polynomials for applications and examples it is more convenient to use the barycentric expression, which is given by

$$\mathcal{L}_{-n,n-1}(z) = \frac{\sum_{j=1}^{2n} \frac{w_j}{z-\alpha_j} u_j}{\sum_{j=1}^{2n} \frac{w_j}{z-\alpha_j}}, \tag{8}$$

with $w_j = \frac{\alpha_j^n}{W'_{2n}(\alpha_j)}$, (see [17]).

This last expression has some advantages due to its numerical stability in the sense established in [18]. In this article author claims literally:

The Lagrange representation of the interpolating polynomial can be rewritten in two more computationally attractive forms: a modified Lagrange form and a barycentric form. We give an error analysis of the evaluation of the interpolating polynomial using these two forms. The modified Lagrange formula is shown to be backward stable. The barycentric formula has a less favourable error analysis, but is forward stable for any set of interpolating points with a small Lebesgue constant. Therefore the barycentric formula can be significantly less accurate than the modified Lagrange formula only for a poor choice of interpolating points.

So with a good Lebesgue constant (see next Theorem 1) we have good accuracy, at least as good as the intensively used Lagrange interpolation on the Chebyshev nodal systems.

Following [10] we can obtain the Lebesgue constant, (see [19]), and the convergence of this interpolatory process. Notice that this is a novelty result for our general nodal systems satisfying property (1), although the techniques that we use are the same as in [10].

Theorem 1. *There exists a positive constant $L > 0$ such that for every function F bounded on $\mathbb{T}$ it holds that*

$$|\mathcal{L}_{-n,n-1}(F,z)| \leq L \, \| F \|_\infty \log n,$$

for every $z \in \mathbb{T}$, where $\| \|_\infty$ denotes the supremum norm on $\mathbb{T}$.

Proof. Let z be an arbitrary point of $\mathbb{T}$ and assume that z is not a nodal point. Then, if we continue assuming the even case, from (7) we get

$$|\mathcal{L}_{-n,n-1}(z)| \leq \sum_{j=1}^{2n} \left| \frac{W_{2n}(z)F(\alpha_j)}{W'_{2n}(\alpha_j)(z-\alpha_j)} \right|,$$

and by our hypothesis about F and by Proposition 3 we have

$$|\mathcal{L}_{-n,n-1}(z)| \leq \frac{\| F \|_\infty}{2nC} \sum_{j=1}^{2n} \left| \frac{W_{2n}(z)}{z-\alpha_j} \right|.$$

If we assume that the nodal points closest to z are α_1 and α_{2n} then by applying (1) we obtain that for $j > 1$ it holds

$$\widehat{z-\alpha_j} > (j-1)\left(\frac{2\pi}{2n} + \mathcal{O}\left(\frac{1}{4n^2}\right)\right).$$

Thus, by using (3) we obtain

$$\frac{1}{|z-\alpha_j|} < \frac{\pi}{2} \frac{2n}{(j-1)} \frac{1}{(2\pi + \mathcal{O}(\frac{1}{2n}))} = \frac{nE}{j-1},$$

for some positive constant E.

Proceeding in the same way we get $z - \widehat{\alpha_{2n-j}} > j(\frac{2\pi}{2n} + \mathcal{O}(\frac{1}{4n^2}))$ and therefore $\frac{1}{|z - \alpha_{2n-j}|} < \frac{nE}{j}$.

We also obtain
$$\left|\frac{W_{2n}(z)}{z - \alpha_1}\right|, \left|\frac{W_{2n}(z)}{z - \alpha_{2n}}\right| < 2nK,$$

for some positive constant K.

Indeed $|W_{2n}(z)| = |W_{2n}(z) - W_{2n}(\alpha_1)| = |W_{2n}(e^{i\theta}) - W_{2n}(e^{i\theta_1})| \leq 2 \max_{z \in \mathbb{T}} |W'_{2n}(z)||\theta - \theta_1| \leq \max_{z \in \mathbb{T}} |W'_{2n}(z)|\pi|z - \alpha_1| \leq 2nK|z - \alpha_1|$.

Hence
$$|\mathcal{L}_{-n,n-1}(z)| \leq \frac{\|F\|_\infty}{2nC} \left(\left|\frac{W_{2n}(z)}{z-\alpha_1}\right| + \sum_{j=2}^{n} \left|\frac{W_{2n}(z)}{z-\alpha_j}\right| + \sum_{j=1}^{n-1} \left|\frac{W_{2n}(z)}{z-\alpha_{2n-j}}\right| + \left|\frac{W_{2n}(z)}{z-\alpha_{2n}}\right| \right) \leq$$

$$\frac{\|F\|_\infty}{2nC} \left(4nK + 2\sum_{j=2}^{n} \frac{2e^A nE}{j-1} \right) = \frac{2\|F\|_\infty}{C} \left(K + \sum_{j=2}^{n} \frac{e^A E}{j-1} \right) \leq 2 \|F\|_\infty P(1 + H_{n-1,1}),$$

with $H_{n-1,1}$ the harmonic number equal to $\sum_{j=1}^{n-1} \frac{1}{j}$ and P a positive constant. □

Remark 2. *When the values of $F(\alpha_i)$ are affected by any type of error, which we can suppose is bounded by some $\epsilon > 0$, then the previous result ensures us, taking into account the linearity of the interpolation process, that the final result is affected by an error bounded by $L \log(n) \epsilon$, that is, it is at least so good as the intensively used Lagrange interpolation on the Chebyshev nodal systems.*

For applying the interpolation it could be very useful the following results concerning the convergence and the rate of convergence for smooth continuous functions (see [10,12]).

Theorem 2. *(i) Let $F(z)$ be a function defined on $\mathbb{T}$. If F is continuous with modulus of continuity $\omega(F, \delta) = o(|\log \delta|^{-1})$, then $\mathcal{L}_{-n,n-1}(F, z)$ converges uniformly to F on $\mathbb{T}$.*

(ii) Let $F(z)$ be a function defined on $\mathbb{T}$. If $F(z) = \sum_{-\infty}^{\infty} A_k z^k$ with $|A_k| \leq K \frac{1}{|k|^c}$ for $k \neq 0$, with $c > 1$ then $\mathcal{L}_{-n,n-1}(F, z)$ uniformly converges to F on $\mathbb{T}$ and the rate of convergence is $\mathcal{O}\left(\frac{\log n}{n^{c-1}}\right)$.

(iii) If $F(z)$ is an analytic function in an open annulus containing $\mathbb{T}$, then $\mathcal{L}_{-n,n-1}(F, z)$ uniformly converges to F on $\mathbb{T}$. Besides, the rate of convergence is geometric.

Proof. The results are consequence of the preceding Theorem 1 and they are also based on the properties satisfied by our nodal systems. Thus one can obtain these results following the same steps as in the proof of Theorems 3 and 4 in [10], where one can see the details. □

The Case of the Bounded Interval

We recall that the Lagrange interpolation polynomial $\ell_{n-1}(x)$ related to a nodal system $\{x_j\}_{j=1}^n$ in $[-1, 1]$ and satisfying the conditions $\{v_j\}_{j=1}^n$ is given by

$$\ell_{n-1}(x) = \sum_{j=1}^{n} \frac{w_n(x)}{w'_n(x_j)(x - x_j)} v_j,$$

where $w_n(x) = \prod_{j=1}^{n} (x - x_j)$. When $v_j = f(x_j)$ for a function f, we write $\ell_{n-1}(f, x)$.

In this subsection we consider the nodal polynomial $w_n(x) = \prod_{j=1}^{n}(x - x_j)$ with $\{x_j\}_{j=1}^{n} \subset [-1,1]$ and numbered in the following way: $1 \geq x_1 > x_2 > \cdots > x_{n-1} > x_n \geq -1$.

We also assume that the nodes satisfy the following separation property:

There exists a positive constant A such that for $n > \dfrac{A}{\pi}$

(i) If $x_1 = 1$ and $x_n = -1$ then $\arccos x_{j+1} - \arccos x_j = \dfrac{\pi}{n} + \dfrac{a(j)}{n^2}$, with $|a(j)| \leq A$, $\forall j = 1, \cdots, n-1$.

(ii) If $x_1 < 1$ and $x_n = -1$ then $\arccos x_{j+1} - \arccos x_j = \dfrac{\pi}{n} + \dfrac{a(j)}{n^2}$, with $|a(j)| \leq A$, $\forall j = 1, \cdots, n-1$, and $2\arccos x_1 = \dfrac{\pi}{n} + \dfrac{a(0)}{n^2}$, with $|a(0)| \leq A$.

(iii) If $x_1 = 1$ and $x_n > -1$ then $\arccos x_{j+1} - \arccos x_j = \dfrac{\pi}{n} + \dfrac{a(j)}{n^2}$, with $|a(j)| \leq A$, $\forall j = 1, \cdots, n-1$, and $2(\pi - \arccos x_n) = \dfrac{\pi}{n} + \dfrac{a(n)}{n^2}$, with $|a(n)| \leq A$.

(iv) If $x_1 < 1$ and $x_n > -1$ then $\arccos x_{j+1} - \arccos x_j = \dfrac{\pi}{n} + \dfrac{a(j)}{n^2}$, with $|a(j)| \leq A$, $\forall j = 1, \cdots, n-1$, and $2\arccos x_1 = \dfrac{\pi}{n} + \dfrac{a(0)}{n^2}$, with $|a(0)| \leq A$, and $2(\pi - \arccos x_n) = \dfrac{\pi}{n} + \dfrac{a(n)}{n^2}$, with $|a(n)| \leq A$.

Under the above assumptions we obtain the following results about the convergence and the rate of convergence for the interpolation polynomials with these nodal systems.

Theorem 3. *If f is a continuous function on $[-1,1]$ and $\omega(f, \delta) = o(|\log \delta|^{-1})$, then the interpolation polynomial $\ell_{n-1}(f, x)$ fulfilling the interpolation conditions*

$$\ell_{n-1}(f, x_j) = f(x_j) \text{ for } j = 1, \cdots, n \qquad (9)$$

converges uniformly to f on $[-1,1]$.

Proof. Let us define a continuous function F on $\mathbb{T}$ by $F(z) = F(\bar{z}) = f(x)$ with $x = \dfrac{z + \dfrac{1}{z}}{2}$. Then it is clear that its modulus of continuity satisfies

$$\omega(F, \delta) = \sup_{z_1, z_2 \in \mathbb{T}, |z_1 - z_2| < \delta} |F(z_1) - F(z_2)| \leq \sup_{x_1, x_2 \in [-1,1], |x_1 - x_2| < \delta} |f(x_1) - f(x_2)| = \omega(f, \delta).$$

To fix ideas we assume that $x_1 \neq 1$ and $x_n \neq -1$, that is, case (iv). By applying Szegő transformation $w_n\left(\dfrac{z + \dfrac{1}{z}}{2}\right) = \dfrac{1}{2^n z^n} \prod_{j=1}^{n}(z - \alpha_j) \prod_{j=1}^{n}(z - \overline{\alpha_j})$, where $\dfrac{\alpha_j + \overline{\alpha_j}}{2} = x_j$, that is, $\alpha_j = e^{i\theta_j}$ with $\theta_j = \arccos x_j$. Hence we consider the nodal polynomial $W_{2n}(z) = 2^n z^n w_n\left(\dfrac{z + \dfrac{1}{z}}{2}\right) = \prod_{j=1}^{n}(z - \alpha_j) \prod_{j=1}^{n}(z - \overline{\alpha_j})$.

Now our nodal system is constituted by $\{\alpha_j\}_{j=1}^{n} \cup \{\overline{\alpha_j}\}_{j=1}^{n}$ and the arguments are $\{\theta_j\}_{j=1}^{n} \cup \{-\theta_j\}_{j=1}^{n}$. If we renumber the arguments in such a way that $-\theta_n = \theta_{n+1}, \cdots, -\theta_1 = \theta_{2n}$, then it holds that

$$\theta_{j+1} - \theta_j = \dfrac{2\pi}{2n} + \dfrac{A(j)}{n^2},$$

with $|A(j)| \leq A$ for $j = 1, \cdots, 2n$ and $\theta_{2n+1} = \theta_1$.

Let $\mathcal{L}_{-n,n-1}(F,z)$ be the Lagrange interpolation polynomial satisfying the conditions

$$\mathcal{L}_{-n,n-1}(F,\alpha_j) = \mathcal{L}_{-n,n-1}(F,\overline{\alpha_j}) = f(x_j), \text{ for } j = 1,\cdots,n.$$

Since F satisfies the hypothesis of Theorem 2 (i) then $\mathcal{L}_{-n,n-1}(F,z)$ converges uniformly to F on $\mathbb{T}$.

Analyzing the expression of $\mathcal{L}_{-n,n-1}(F,z)$ and by taking into account that $W_{2n}(z)$ as well as $W'_{2n}(z)$ have real coefficients we get that $\mathcal{L}_{-n,n-1}(F,z)$ has real coefficients. Since it is clear that $\mathcal{L}_{-n,n-1}(F,\frac{1}{z}) \in \Lambda_{-(n-1),n}$ satisfies the same interpolation conditions, now we consider the algebraic polynomial $\frac{1}{2}\left(\mathcal{L}_{-n,n-1}(F,z) + \mathcal{L}_{-n,n-1}(F,\frac{1}{z})\right)$, which has real coefficients and satisfies the interpolation conditions (9). Since the polynomial satisfying (9) is uniquely determined, then $\frac{1}{2}\left(\mathcal{L}_{-n,n-1}(F,z) + \mathcal{L}_{-n,n-1}(F,\frac{1}{z})\right) = \ell_{n-1}(f,x)$ and it converges uniformly to f on $[-1,1]$.

When $x_1 = 1$ (case (iii)), or $x_n = -1$ (case (ii)), or $x_1 = 1$ and $x_n = -1$ (case (i)), one proceeds in a similar way and the auxiliary nodal polynomials are given by $W_{2n-1}(z) = \frac{2^n z^n}{z-1} w_n(\frac{z+\frac{1}{z}}{2}) = (z-1)\prod_{j=2}^{n}(z-\alpha_j)\prod_{j=2}^{n}(z-\overline{\alpha_j})$ or $W_{2n-1}(z) = \frac{2^n z^n}{z+1} w_n(\frac{z+\frac{1}{z}}{2}) = (z+1)\prod_{j=1}^{n-1}(z-\alpha_j)\prod_{j=1}^{n-1}(z-\overline{\alpha_j})$ or $W_{2n-2}(z) = \frac{2^n z^n}{(z-1)(z+1)} w_n(\frac{z+\frac{1}{z}}{2}) = (z-1)(z+1)\prod_{j=2}^{n-1}(z-\alpha_j)\prod_{j=2}^{n-1}(z-\overline{\alpha_j})$, respectively. Notice that in the four cases the nodal polynomials have real coefficients. □

Notice that some well known results related to Lagrange interpolation with Chebyshev and Chebyshev extended nodes are particular cases of the above theorem (see [9,20]).

Notice that from the proof of the above theorem and by applying Theorem 1 we can obtain an analogous bound, the Lebesgue constant, for this interpolatory process, that is, there exists a positive constant M such that

$$|\ell_{n-1}(f,x)| \leq M \|f\|_\infty \log n. \tag{10}$$

In order to obtain information concerning the rate of convergence, first we recall the following result about the expansion of an analytic function in a Jacobi series (see [9,21]). For a more actual version see [12].

Theorem 4. *Let $f(x)$ be analytic on the closed segment $[-1,1]$. The expansion of f in a Jacobi series, $f(x) \sim \sum_{n=0}^{\infty} a_n P_n^{(\alpha,\beta)}(x)$, is convergent in the interior of the greatest ellipse with foci at ± 1, in which f is regular. The expansion is divergent in the exterior of this ellipse and the sum R of the semi-axes of the ellipse of convergence is $R = \liminf \frac{1}{\sqrt[n]{|a_n|}}$.*

Thus, in our conditions we have the following results, which are in concordance with Theorem 2.

Theorem 5. *(i) If f is a function defined on $[-1,1]$ by $f(x) = \sum_{k=0}^{\infty} a_k T_k(x)$, where $T_k(x)$ is the Chebyshev polynomial of degree k and with $|a_k| \leq \frac{K}{k^s}$, with $k \neq 0$, $K > 0$ and $s > 1$, then the Lagrange interpolation polynomial $\ell_{n-1}(f,x)$ converges to f with rate of convergence $\mathcal{O}\left(\frac{\log n}{n^{s-1}}\right)$.*

(ii) If f is analytic on the closed segment $[-1,1]$, then the Lagrange interpolation polynomial $\ell_{n-1}(f,x)$ converges to f with rate of convergence geometric.

Proof. (i) If we decompose f like $f(x) = f_{1,n-1}(x) + f_{2,n-1}(x) = \sum_{k=0}^{n-1} a_k T_k(x) + \sum_{k=n}^{\infty} a_k T_k(x)$, we have that $\ell_{n-1}(f_{1,n-1}, x) = f_{1,n-1}(x)$ and $|f_{2,n-1}(x)| \leq \sum_{k=n}^{\infty} |a_k| \leq K \sum_{k=n}^{\infty} \frac{1}{k^s} < \frac{K}{(s-1)(n-1)^{s-1}}$, where the last inequality follows from the integral criterion. Thus, by applying the Lebesgue constant obtained before in (10), we get $|\ell_{n-1}(f_{2,n-1}, x)| \leq M \frac{K}{(s-1)(n-1)^{s-1}} \log n$. Hence

$$|f(x) - \ell_{n-1}(f, x)| = |f_{1,n-1}(x) - \ell_{n-1}(f_{1,n-1}, x) + f_{2,n-1}(x) - \ell_{n-1}(f_{2,n-1}, x)| =$$
$$|f_{2,n-1}(x) - \ell_{n-1}(f_{2,n-1}, x)| \leq \frac{K}{(s-1)(n-1)^{s-1}}(1 + M \log n) \leq T \frac{\log n}{n^{s-1}},$$

for some $T > 0$.

(ii) Since f is analytic on $[-1, 1]$, then it can be analytically continued to a neighborhood of $[-1, 1]$ in the complex plane. Hence the expansion in Chebyshev series $\sum_{n=0}^{\infty} a_n T_n(x)$ converges to f in the interior of the greatest ellipse with foci at ± 1, known as Bernstein ellipse E_R and the sum R of the semi-axes of the ellipse of convergence is $R = \liminf \frac{1}{\sqrt[n]{|a_n|}}$. Then it holds that $|a_n| \leq P r^n$, for some $0 < r < 1$ and $P > 0$.

Proceeding in the same way as before we have $f(x) = f_{1,n-1}(x) + f_{2,n-1}(x) = \sum_{k=0}^{n-1} a_k T_k(x) + \sum_{k=n}^{\infty} a_k T_k(x)$, $\ell_{n-1}(f_{1,n-1}, x) = f_{1,n-1}(x)$ and $|f_{2,n-1}(x)| \leq \sum_{k=n}^{\infty} |a_k| \leq P \frac{r^n}{1-r}$. Thus, by applying the Lebesgue constant we get $|\ell_{n-1}(f_{2,n-1}, x)| \leq MP \frac{r^n}{1-r} \log n$ and therefore

$$|f(x) - \ell_{n-1}(f, x)| = |f_{2,n-1}(x) - \ell_{n-1}(f_{2,n-1}, x)| \leq Qr^n \log n \leq Qr_1^n,$$

for some $Q > 0$ and $0 < r < r_1 < 1$. □

4. Numerical Examples

We have carried out different numerical experiments to visualize the main contributions of this article. The first examples correspond to the three cases of Theorem 2 and in all of them we work in the following way:

1. We construct the nodal systems in a quite random way. We consider four arcs or sections in the unit circumference $\mathbb{T}$. The first one begins in $\alpha_1 = 1$ and its $\frac{n}{4}$ nodes are constructed in counter clockwise sense separated by an angular length $\frac{2\pi}{n} + \epsilon$, where the ϵ are random errors determined by using the uniform distribution in $[\frac{A}{n^2} 2\pi, \frac{2A}{n^2} 2\pi]$. The fourth section begins in $\alpha_1 = 1$ and its $\frac{n}{4}$ nodes are constructed in clockwise sense with arcs of angular length $\frac{2\pi}{n} + \epsilon$, where the ϵ are random errors determined by using the uniform distribution in $[\frac{A}{n^2} 2\pi, \frac{2A}{n^2} 2\pi]$. The second section begins after the first one and its $\frac{n}{4}$ nodes are constructed in counter clockwise sense with arcs of angular length $\frac{2\pi}{n} + \epsilon$, where the ϵ are random errors determined by using the uniform distribution in $[-\frac{2A}{n^2} 2\pi, -\frac{A}{n^2} 2\pi]$. Finally, in the third section the arcs between the nodes are all equal.

Obviously we use different values of n and we must remark that we obtain always the same results, really we must say similar results because due to our random choice we never have the same nodal system.

2. We consider a test function $F(z)$, that we detail in each example, and we always plot $F(z)$ in black.
3. We consider the Lagrange interpolation polynomial $\mathcal{L}_{-n,n-1}(F,)$, which interpolates the test function at the nodal system. We always plot $\Re(\mathcal{L}_{-n,n-1}(F,))$ in red.

These examples are devoted to visualize the items (i), (ii) and (iii) respectively of Theorem 2.

Example 1. *In this example we work with $F(z) = 1 + 20\left(\dfrac{z+z^{-1}}{2}\right)\sin\left(\dfrac{2}{z+z^{-1}}\right)$ for $z \in \mathbb{T}$, which satisfies the hypotheses of Theorem 2 (i). We take $n = 1000$, $A = 2$ and we use (8) to obtain $\mathcal{L}_{-n,n-1}(F,)$.*

We represent the function $F(e^{i\theta})$ which takes real values and, as we have said, the real part of the interpolation polynomial. Notice that due to its variability, F is a quite difficult function to interpolate. Indeed, it is easy to check that $F(e^{i\theta})$ is not differentiable at $\dfrac{\pi}{2}$.

We present in Figure 2 two graphics. On the left we have a general panoramic of the interpolation along $\mathbb{T}$ and we have added the interpolation points in green. We must point up that the interpolatory process is successful where the function has no variability. However, we have an unsuccessful situation where the function has great variability.

In the graphic on the right we have a detailed situation between 1.2 and 2, that is near $\dfrac{\pi}{2}$, which can help us to understand the problem. According to the theory presented, we must increase the number of nodes to obtain better results in this region.

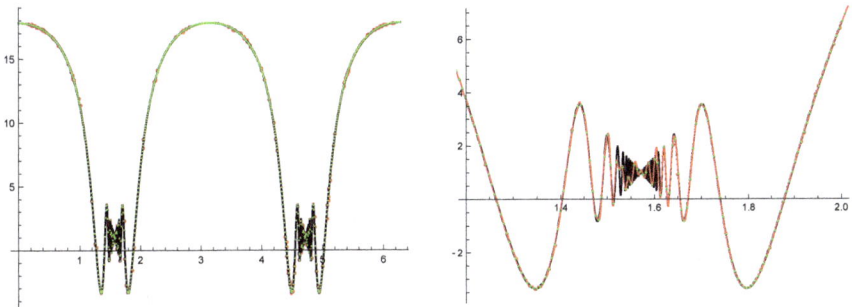

Figure 2. $F(z)$ and $\Re(\mathcal{L}_{-n,n-1}(F,z))$ with F(z)= $1 + 20(\dfrac{z+z^{-1}}{2})\sin(\dfrac{2}{z+z^{-1}}), z = e^{i\theta}, \theta \in [0, 2\pi]$, $\theta \in [1.2, 2]$ and $n = 1000$.

Example 2. *Now we consider the function defined on $\mathbb{T}$ by $F(z) = \sum_{k=1}^{\infty} \dfrac{1}{k^6}(z^k + z^{-k})$, which satisifies the hypotheses of Theorem 2. In the next Figure 3 we plot on the left $F(e^{i\theta})$ and $\Re(\mathcal{L}_{-n,n-1}(F,e^{i\theta}))$ for $\theta \in [0, 2\pi]$ and $n = 60$. Notice that they are indistinguishable. On the right we plot the errors given by $\Re(\mathcal{L}_{-n,n-1}(F,e^{i\theta})) - F(e^{i\theta})$ with $\theta \in [0, 2\pi]$. We point out that the errors are less or equal than 2×10^{-8}.*

In the next example we also construct an alternative interpolation polynomial based on the equispaced nodal system on $\mathbb{T}$, but using the values of the function on our nodal system. We do this because a natural criticism to our method could be that with errors as $\mathcal{O}(1/n^2)$ we can be so close to the equispaced nodal system to accept this approximation. We denote by $\mathcal{A}_{-n,n-1}(F,)$ this alternative interpolation polynomial.

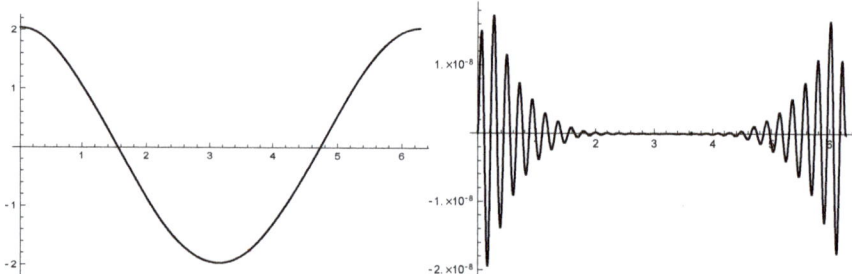

Figure 3. $F(z)$ and $\Re(\mathcal{L}_{-n,n-1}(F,z))$ and $\Re(\mathcal{L}_{-n,n-1}(F,z)) - F(z)$ with $F(z) = \sum_{k=1}^{\infty} \frac{1}{k^6}(z^k + z^{-k})$, $z = e^{i\theta}$, $\theta \in [0, 2\pi]$ and $n = 60$.

Example 3. In this example we take $F(z) = e^z$, $n = 24$, $A = 2$ and we use (8) to obtain the interpolation polynomials $\mathcal{L}_{-n,n-1}(F,)$ and $\mathcal{A}_{-n,n-1}(F,)$. Taking into account that F is analytic we know that F and $\mathcal{L}_{-n,n-1}(F,)$ must be close. In Figure 4 we plot $\Re(F)$ in black, $\Re(\mathcal{L}_{-n,n-1}(F,))$ in red and $\Re(\mathcal{A}_{-n,n-1}(F,))$ in green for $z = e^{i\theta}$ with $\theta \in [0, 2\pi]$. On the left hand side we have a global vision with $\theta \in [0, 2\pi]$ and we can observe that $\Re(F)$ and $\Re(\mathcal{L}_{-n,n-1}(F,))$ are indistinguishable; in fact for this example the maximum error was 3×10^{-9}.

Although $\Re(\mathcal{A}_{-n,n-1}(F,))$ has a similar shape, see that it can drive us to catastrophic errors. On the right hand side we present a detail of the previous one, which give us an idea of the error. Notice that in general we cannot know the details of the nodal distribution.

We have done the same with the imaginary part and we obtain the same results.

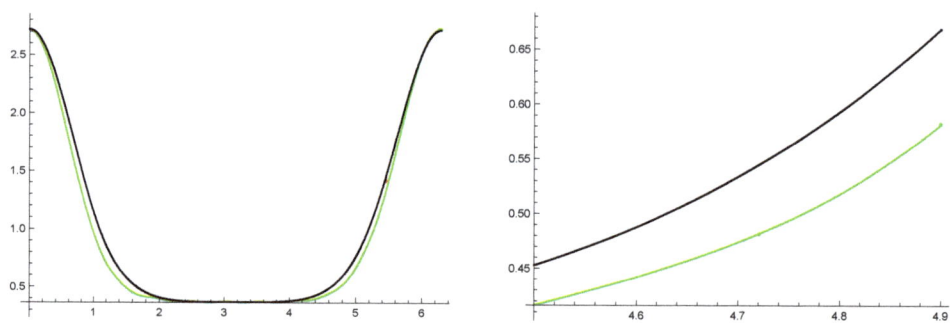

Figure 4. $\Re(F(z))$, $\Re(\mathcal{L}_{-n,n-1}(F,z))$ and $\Re(\mathcal{A}_{-n,n-1}(F,z))$ with $F(z) = e^z$, $z = e^{i\theta}$, $\theta \in [0, 2\pi]$, $\theta \in [4.5, 4.9]$ and $n = 24$.

Example 4. Finally we choose $F(z) = \chi_S(z)$ defined on $\mathbb{T}$ as the characteristic function of the superior arc S of $\mathbb{T}$, we take $n = 2000$, $A = 2$ and we use expression (8) to obtain $\mathcal{L}_{-n,n-1}(F,)$. We know the behavior when the nodal system is related to para-orthogonal polynomials with respect to an analytic positive measure (see [10]), but we do not have a theory about the behavior of $\mathcal{L}_{-n,n-1}(F,)$ in our situation. We plot the results in Figure 5. Notice that the basic ideas of the Gibbs–Wilbraham phenomenon are present in this graphic, that is, the convergence of the interpolator to the function in regions which are far enough from the discontinuities and a heavy oscillation near the discontinuities. A representation of the oscillation and its amplitude, maybe, deserves a detailed study.

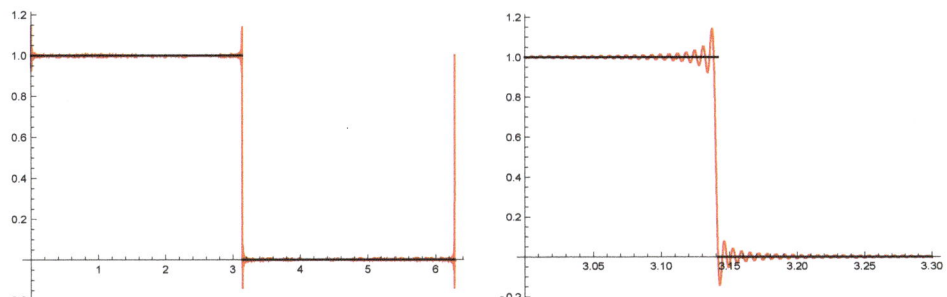

Figure 5. $F(e^{i\theta})$ and $\Re(\mathcal{L}_{-n,n-1}(F,e^{i\theta}))$ with $F(z) = \chi_S(z), z = e^{i\theta}, \theta \in [0,2\pi], \theta \in [3,3.3]$ and $n = 2000$.

5. Discussion

Usually, the nodal systems used for interpolation problems are strongly connected with measures on the bounded interval and on the unit circle and their corresponding orthogonal or paraorthogonal polynomials. We must point out that these choices are very suitable to construct the whole theory but in some numerical applicatons it is possible that the nodal systems do not satisfy this requisite. So, the starting point of the paper is a distribution for the nodes that can be obtained through a perturbation of the uniform distribution and, in particular of the roots of the unity, and which is more general than that related to measures and orthogonality.

The results of this article contribute to elaborate a theory over these type of nodal systems, as well as to the Lagrange interpolation theory based on these interpolatory arrays. Moreover, a theory about the rate of convergence for some types of smooth functions is given. Finally, we translate the results to perturbed Chevyshev nodal systems and to Lagrange interpolation on the bounded interval.

We think that this research could be of interest for some mechanical models that generate these types of nodal systems. As an example we consider the next problem.

Let us suppose that we are studying a equatorial characteristic $F(e^{i\theta})$ of a planet which depends on the angle θ and we have a theory which establishes that $F(e^{i\theta})$ is an analytic function. We observe the phenomenon using an observatory in the boundary of a spatial station in an elliptic orbit of period T which rotates over itself with period T_1 (with $T = n\,T_1$ and n large enough). Moreover, we take our observations when the center of the planet, the observatory and the center of the station are aligned. We can translate the problem thinking that the planet is our Sun, the spatial station is our Earth and the observatory is our city. So the time between observations is the equivalent of a solar day. It is well known that the duration of a solar day is not constant (see https://en.wikipedia.org/wiki/Equation_of_time for a brief introduction about the so called **Equation of time**), in our case have a little oscillation on T_1 and our observations are taken on a nodal system which satisfies (1). Notice that in this case we do not have a equispaced distribution nor the support of the theory of Orthogonal Polynomials. Therefore, before this paper we did not know how to use our data to reconstruct $F(e^{i\theta})$ and after this paper we can be confident about the use of Lagrange interpolation.

Some future research directions could be the study of other types of interpolation on the unit circle and on the bounded interval by using these general interpolatory arrays; as well as to study the correspondig Gibbs–Wilbraham phenomena.

6. Materials and Methods

The experiments given in the section Numerical examples were obtained by using personal codes elaborated with Mathematica® 12 (Wolfram Research Europe Ltd, Long Hanborough Oxfordshire, United Kingdom). These programs to obtain the nodal points and to compute the interpolation polynomials as well as the plots of the test functions and their interpolators are available at the public

repository https://www.dropbox.com/sh/0cx9chq3jfzov2w/AAA_SvL2i7HlC7ChMGpuG-Ata?dl=0
There one can find the program related to Example 2. To obtain the other examples some minor changes must be done.

Author Contributions: Conceptualization, E.B., A.C. (Alicia Cachafeiro), A.C. (Alberto Castejón) and J.M.G.-A.; Investigation, E.B., A.C. (Alicia Cachafeiro), A.C. (Alberto Castejón) and J.M.G.-A.; Software, E.B., A.C. (Alicia Cachafeiro), A.C. (Alberto Castejón) and J.M.G.-A.; Writing—original draft, E.B., A.C. (Alicia Cachafeiro), A.C. (Alberto Castejón) and J.M.G.-A. All authors have read and agreed to the published version of the manuscript.

Funding: This research received no external funding.

Conflicts of Interest: The authors declare no conflict of interest.

References

1. Szili, L.; Vértesi, P. On the theorem of Géza Grünwald and József Marcinkiewicz. *Banach Cent. Publ.* **2011**, *95*, 251–259. [CrossRef]
2. Erdös, P.; Vértesi, P. On the almost everywhere divergence of Lagrange interpolation of polynomials for arbitrary systems of nodes. *Acta Math. Acad. Sci. Hung.* **1980**, *36*, 71–89. [CrossRef]
3. Trefethen, L.N. Six Myths of Polynomial Interpolation and Quadrature. *Math. Today (Southend-on-Sea)* **2011**, *47*, 184–188.
4. Grünwald, G. On the theory of interpolation. *Acta Math.* **1942**, *75*, 219–245. [CrossRef]
5. Erdös, P.; Turán, P. On Interpolation II: On the Distribution of the Fundamental Points of Lagrange and Hermite Interpolation. *Ann. Math.* **1938**, *9*, 703–724. [CrossRef]
6. Walsh, J.L. *Interpolation and Approximation by Rational Functions in the Complex Plane*, 5th ed.; American Mathematical Society Colloquium Publications: Providence, RI, USA, 1969; Volume 20.
7. Darius, L.; González-Vera, P. Some results about interpolation with nodes on the unit circle. *Indian J. Pure Appl. Math.* **2000**, *31*, 1273–1296.
8. Berriochoa, E.; Cachafeiro, A.; García-Amor, J.M. About nodal systems for Lagrange interpolation on the circle. *J. Appl. Math.* **2012**, *2012*. [CrossRef]
9. Szegő, G. *Orthogonal Polynomials*, 4th ed.; American Mathematical Society Colloquium Publications: Providence, RI, USA, 1975; Volume 23.
10. Berriochoa, E.; Cachafeiro, A.; García-Amor, J.M. Gibbs–Wilbraham phenomenon on Lagrange interpolation based on analytic weights on the unit circle. *J. Comput. Appl. Math.* **2020**, *365*, 112376. [CrossRef]
11. Chui, C.K.; Shen, X.-C.; Zhong, L. On Lagrange interpolation at disturbed roots of unity. *Trans. Am. Math. Soc.* **1993**, *336*, 817–830. [CrossRef]
12. Trefethen, L.N. *Approximation Theory and Approximation Practice*; Society for Industrial and Applied Mathematics (SIAM): Philadelphia, PA, USA, 2013.
13. Milovanovic, G.V.; Mitrinovic, D.S.; Rassias, T.M. *Topics in Polynomials: Extremal Problems, Inequalities, Zeros*; World Scientific Publishing Co.: Singapore, 1994.
14. Berriochoa, E.; Cachafeiro, A.; Martínez, E. About measures and nodal systems for which the Hermite interpolants uniformly converge to continuous functions on the circle and interval. *Appl. Math. Comput.* **2012**, *218*, 4813–4824. [CrossRef]
15. Berriochoa, E.; Cachafeiro, A.; García-Amor, J.M. Gibbs–Wilbraham oscillation related to an Hermite interpolation problem on the unit circle. *J. Comput. Appl. Math.* **2018**, *344*, 657–675. [CrossRef]
16. Simon, B. Fine structure of the zeros of orthogonal polynomials, I. A tale of two pictures. *Electron. Trans. Numer. Anal.* **2006**, *25*, 328–368.
17. Corless, R.M.; Fillion, N. *A Graduate Introduction to Numerical Methods*; Springer: New York, NY, USA, 2013.
18. Higham, N.J. The numerical stability of barycentric Lagrange interpolation. *IMA J. Numer. Anal.* **2004**, *24*, 547–556. [CrossRef]
19. Berrut, J.-P.; Mittelmann, H.D. Lebesgue constant minimizing linear rational interpolation of continuous functions over the interval. *Comput. Math. Appl.* **1997**, *33*, 77–86. [CrossRef]

20. Szabados, J.; Vértesi, P. *Interpolation of Functions*; World Scientific: Singapore, 1990.
21. Rivlin, T.H. *The Chebyshev Polynomials*; John Wiley & Sons: New York, NY, USA, 1974.

 © 2020 by the authors. Licensee MDPI, Basel, Switzerland. This article is an open access article distributed under the terms and conditions of the Creative Commons Attribution (CC BY) license (http://creativecommons.org/licenses/by/4.0/).

Article

A Functional Data Analysis Approach for the Detection of Air Pollution Episodes and Outliers: A Case Study in Dublin, Ireland

Javier Martínez Torres [1],*, Jorge Pastor Pérez [2], Joaquín Sancho Val [3], Aonghus McNabola [4], Miguel Martínez Comesaña [5] and John Gallagher [4]

1. Department of Applied Mathematics I. Telecommunications Engineering School, University of Vigo, 36310 Vigo (Pontevedra), Spain
2. Centro de Evaluación, Formación y Calidad de Aragón, 50018 Zaragoza, Spain; jjpastor@aragon.es
3. Centro Universitario de la Defensa. Academia General Militar, 50090 Zaragoza, Spain; jsanchov@unizar.es
4. Department of Civil, Structural and Environmental Engineering, Trinity College Dublin, University of Dublin, Dublin D02 PN40, Ireland; amcnabol@tcd.ie (A.M.); J.Gallagher@tcd.ie (J.G)
5. Escuela de Ingeniería Industrial, University of Vigo, 36310 Vigo (Pontevedra), Spain; migmartinez@uvigo.es
* Correspondence: javmartinez@uvigo.es

Received: 14 January 2020; Accepted: 4 February 2020; Published: 10 February 2020

Abstract: Ground level concentrations of nitrogen oxide (NOx) can act as an indicator of air quality in the urban environment. In cities with relatively good air quality, and where NOx concentrations rarely exceed legal limits, adverse health effects on the population may still occur. Therefore, detecting small deviations in air quality and deriving methods of controlling air pollution are challenging. This study presents different data analytical methods which can be used to monitor and effectively evaluate policies or measures to reduce nitrogen oxide (NOx) emissions through the detection of pollution episodes and the removal of outliers. This method helps to identify the sources of pollution more effectively, and enhances the value of monitoring data and exceedances of limit values. It will detect outliers, changes and trend deviations in NO_2 concentrations at ground level, and consists of four main steps: classical statistical description techniques, statistical process control techniques, functional analysis and a functional control process. To demonstrate the effectiveness of the outlier detection methodology proposed, it was applied to a complete one-year NO_2 dataset for a sub-urban site in Dublin, Ireland in 2013. The findings demonstrate how the functional data approach improves the classical techniques for detecting outliers, and in addition, how this new methodology can facilitate a more thorough approach to defining effect air pollution control measures.

Keywords: air pollution; functional data analysis; non-normal data; statistical process control; outlier

1. Introduction

Nowadays, most cities have an increasing environmental problem related to air pollution [1–4]. This specific pollution is a continuing threat to human health and welfare, with a range of different sources generating different pollutants which have distinct health effects on urban populations [5–7]. Detailed air quality monitoring data for pollutants, such as carbon monoxide (CO), nitrogen oxides (NO and NO_2), sulphur dioxide (SO_2), ozone (O_3) and particulate matter (PM_{10} and $PM_{2.5}$), are becoming more important because of the health problems said pollutants can cause in living beings [6]. The measurements of pollutants provide real-time data to inform the public and provide a mechanism of alerting local residents of a possible hazard. In particular, pollutant sources from traffic emissions, such as NOx, which represents a combination of nitrogen oxide (NO) and nitrogen dioxide (NO_2), are typically emitted at ground level from vehicles and are associated with health-related problems [8].

Despite a reduction in emissions from the transport sector, an increasing trend in NO_2 concentrations has been observed in a number of different European countries; for example, the United Kingdom and Ireland [9,10]. Therefore, meeting the standards and air quality guidelines by European and national environmental agencies for pollutants such as NO_2 is becoming more challenging [9], as exceedances of pollutant concentrations can lead to short-term, chronic human health problems [11].

On the other hand, it is understandable that occasional values in polluted air samples behave as outliers in an urban environmental database. They can be classified as local outliers [12,13] or global outliers. Unlike global outliers, local outliers can be detected by comparison with near neighbours. For the purpose of air pollution studies in urban areas, global outliers that deviate from the guide values indicate that there may be a significant source of pollution. Observations which are not excessively high but are different from neighbouring values may also contain information on unusual processes such as pollution. Outliers may merely be noisy observations, or alternatively, they may indicate atypical behaviour in the system. These abnormal values are very important and may lead to useful information or significant discoveries, but also contribute to the selection of the most suitable mitigation techniques or measures [14].

Different techniques of functional data analysis (FDA) have been used in vectorial problems. This new methodology appeared due to the inefficiency of the classical data mining techniques treating vector data [15]. FDA is applicable in a multitude of fields, such as environmental [16–19] and medical research [20], and is applicable for sensors [21,22] and industrial methods [23,24]. The functional model is based on two ideas that make it unique: it takes into account the time correlation structure of the data and leads to a global view of the problem through curves analysis instead individual observations. This analysis is focused on the comparison of the curves using the functional depths, a variable that measures the centrality of a given curve within a set of curves [25]. Functional depth has already been used in several environmental problems [26,27].

The aspiration of this research is to create a model to detect air pollution episodes and identify outliers in gaseous emissions, and to validate this method using real world data from a suburban air quality monitoring site in Dublin, Ireland. Although many methods are known to identify outliers (from the classical Grubbs test [28] to a test proposed in 2019 by [29]), they are all based on the vector approach. This study was carried out, on one hand, with conventional methods, and on the other hand, with a functional approach; a comparative study between the two methodologies is presented. Each method will be presented and the findings will outline the most effective method for detecting outliers in air pollution monitoring data to enhance its capacity for informing new measures to improve local air quality.

2. Methods

2.1. Case Study—A Sub-Urban Air Quality Monitoring Station in Dublin, Ireland

Ireland has a range of air quality monitoring stations across the country, which are part of the national ambient air quality monitoring programme (AAMP). The data collated from these monitoring sites are used to inform on air quality at the local and national levels, and are being used for forecast modelling. The Blanchardstown sub-urban site is one of the 17 national sites, managed by the Environmental Protection Agency (EPA), which monitors NO_2 and is classified as a suburban monitoring site. It is located to the west of Dublin city centre in Ireland [30]. The Blanchardstown air quality monitoring station was selected, as it provided continuous, high-resolution NO_2 data emissions over a 1-year period. Its location is adjacent to the major arterial carriageway around Dublin city centre, and as a monitoring location, is therefore affected by traffic emissions. In this manner, NO_2 hourly data was collected throughout 2013, with 96% of data capture and availability from [31], and the information needed about weather conditions in Dublin in 2013 was obtained on [32].

EU legislation for NO_2 limit values (2008 CAFE Directive and S.I. number 180 of 2011) align with the World Health Organisation (WHO) guidelines, with 1-hour and 1-year limit values of 200 $\mu g/m^3$

and 40 µg/m^3 respectively. In circumstances in which the hourly value is exceeded on three consecutive hours, short-term action plans must be implemented by local authorities to mitigate against continued pollution events (limiting traffic flows, restricting construction work, industrial processes, etc.). Despite there being no daily NO$_2$ limit value in the EU and WHO, some countries have set an average daily limit of 100 µg/m^3 (range from 80–150 µg/m^3) [31].

The sources and trends of NO$_2$ emissions over the last 20 years has seen a recent increasing trend due to growth in the transport sector and a recovery since the 2008 economic downturn. As such, Ireland's air quality in relation to NO$_2$ is considered to be deteriorating, as measurement data suggests it may reach limit values and the national emissions ceiling in the coming years.

2.2. Analysis Methods

A range of systems are available to analyse environmental data, such as air quality measurements. These systems can be used to detect uncharacteristic data points by taking into account trends, variations between neighbouring network stations and expected values with respect to the sampling location. An example of these expert systems of data and environmental parameter validation would be the trends analysis throughout R-programming (openair) [33].

With classical analysis, the data are only analysed statically. The proposed methodology includes using a large amount of existing data to extract conclusions. Today, the amount of data that has been stored in environmental databases requires automated analysis techniques. Actually, the analysis methodology presented here is oriented to knowledge discovery in databses (KDD) [34], which provides a complete process for extracting information and also provides a clear methodology for the preparation of data and interpreting the results obtained. KDD involves an iterative and interactive process of searching for models, patterns and parameters that are useful for detection, classification and/or prediction in order to generate knowledge and help in decision making.

2.2.1. Classical Analysis

The classical monitoring strategy for air quality uses individual time series, descriptive statistics, box plots, autocorrelation analysis, etc., to determine if any of the values fall outside of the limits and to analyse trends [35,36]. In general, classical statistical analysis seeks to describe the distribution of a measurable variable (descriptive statistics) and to determine the consistency of a sample drawn from an initial population (inferential statistics). In addition, classical analysis is based on repetition; one must measure properties of objects and try to predict the frequency of occurrence of results when the measuring operation is repeated at random or stochastically.

This type of analysis determines the empirical frequency distribution that yields the absolute or relative frequency of the occurrence of the different possible results of the repeated measurement of a property of an object (discrete case). Instead, if the case is an infinitely repeated and arbitrarily precise measurement and every outcome is diffferent, the relative frequency of a single outcome would not be very instructive; the distribution function is used, which, for every numerical value x of the measured variable, yields the absolute (or relative) frequency of the occurrence of all values smaller than x [37].

2.2.2. Statistical Process Control

By applying statistical process control (SPC) methods to the monitoring of a system, it is possible to detect outliers. This study is concentrated on significantly high and low measurements, even in situations where the values do not exceed the established limit. These methodologies can be used to study individual observations, using individual or average charts.

The dataset should be partitioned into rational subgroups, minimising the probability of large differences between subgroups [38]. The formation of rational subgroups is important, because variation within subgroups can be clustered and the presence of special causes of variability can be easily detected. However, sometimes it is not practical to use rational subgroups; for example, when repeated measurements in the process differ only by laboratory or analytical error. Even when automated inspection is used, because every unit manufactured is analysed.

The rational subgroups represent the way of collecting the data. Usually, they should be gathered so that each of them shows only the inherent variation that is natural for the process (*common cause variation*). Because they contribute to identifying any other source of variation (*special cause variation*) that may badly affect the process, the subgroups should avoid special-cause variation where possible. Moreover, the limits on a control chart, which mark the boundary to identify if a process is too volatile, are calculated on the basis of the variability within each subgroup. For this, only subgroups that reproduce the common cause variation in a process should be selected.

Once the data are correctly structured, a normality test has to be done. If the hypothesis of normality is rejected, there are two possibilities: use modified classical techniques to non-normal distributions [39], or transform the data to normalise the dataset [40]. The second option applies a Box-Cox transformation [41], which smoothes the data structure. The most widely used and known transformation is the Box-Cox transformation, defined as follows:

$$X_j^{(\lambda)} = \begin{cases} \frac{X_j^\lambda - 1}{\lambda}, & \text{if } \lambda \neq 0 \\ log(X_j), & \text{if } \lambda = 0 \end{cases}$$

where λ maximises the profile likelihood function of the data X_j.

A classical analysis process can be divided in two main stages: the learning stage, when a test of normality is performed and atypical measurements are deleted from the data; and the control stage, when the trends are analysed to encounter *out-of-control* situations. At the first one, it is when the centreline (CL) and control limits are defined. Specifically, the CL is defined with the control sample and represents the objective value. In addition, the warning limits are set at a distance of $\pm 2\sigma$ from it, and control limits at $\pm 3\sigma$, with σ being the standard deviation of the process [42].

Shewhart control charts have been the most widely used due to their good performance in detecting large changes in a process. However, because these charts use the most recent samples, they do not efficiently detect small or progressive changes in a process. In this regard, complementary rules are needed; multiple authors have defined different rules to detect specific deviations [43,44] and to complement the initial control rules [45]. The use of these supplementary rules makes Shewhart's control charts more sensitive and leads to an improvement in one's capacity to detect non-random patterns.

A widely used way to quantify the potential of a control chart with supplementary rules is through the average run length (ARL). The ARL, in control charts, is the average number of points that should be analysed before showing an alert warning that the process is not under control. When this occurs, the efficient thing to do would be to detect it as soon as possible. On the contrary, when the process is statistically stable, it would be appropriate to have few false alarms. This term is directly related to a Type I error (also known as α) and a Type II error (also known as β), which also describe the sensitivity of the method, and it is highly related to the number of false alarms. For that reason, it must be contemplated that if the capacity of this methodology to detect out-of-control situations is high, there will be a lot of false alarms [43].

2.2.3. Functional Data Analysis

The functional data analysis works with the observations that come from a continuous random process that is evaluated at discrete points [46]. Starting from vector samples, the dataset will be transformed into a functional sample. The first step is to construct the most appropriate curves from the initial points that come from the discrete values measured in the study. This process, known as smoothing, converts the vector values into continuous functions over time. This structure of data is essential in the air pollution context because it is taking into account all the values in the day as a set. In this way, a day in which NO_2 values are obtained with a lot of variability but which has an average similar to the other days, is not detected from any vectorial approach. Functional analysis would identify these types of days as candidates for outliers. Furthermore, in several similar studies in which they also tried to detect outliers with data from certain gases (see [23,47]), the superiority of functional approaches was demonstrated.

In a situation where the initial observations $x(t_j)$ are contained in a set of n_p points, $t_j \in \mathbb{R}$ represents the time steps and n_p is the number of observations ($j = 1, 2, \ldots, n_p$). They can be watched as the individual values of the function $x(t) \in \mathbb{X} \subset F$, F being a functional space. The estimation of $x(t)$ takes into account a functional space $F = span\{\phi_1, \ldots, \phi_{n_b}\}$, where ϕ_k is the set of basis functions ($k = 1, 2, \ldots, n_b$) and n_b is the number of basis functions required to build a functional sample. There are several types of basis in statistics, but the most used one is the Fourier basis [26,48]. Moreover, with periodic data such as we have in this study, Fourier bases are the most appropriate [49]. Smoothing consists of finding a solution to the regularisation problem [46],

$$\min_{\mathbb{X} \in F} \sum_{j=1}^{n_p} \{z_j - x(t_j)\}^2 + \lambda \Gamma(x) \tag{1}$$

with $z_j = x(t_j) + \epsilon_j$ being the result of observing x at the point t_j; ϵ_j the random noise with zero mean, Γ being a penalising operator focused on obtaining the simplest possible solutions; and λ being a parameter that defines the level of the regularisation. The initial step is the following expansion

$$x(t) = \sum_{k=1}^{n_b} c_k \phi_k(t) \tag{2}$$

where $\{c_k\}_{k=1}^{n_b}$ are coefficients that multiply the selected basis functions. The smoothing problem can be written as follows:

$$\min_c \{(z - \Phi c)^T (z - \Phi c) + \lambda c^T R c\} \tag{3}$$

with a vector of observations $z = (z_1, \ldots, z_{n_p})^T$; a vector of coefficients of the expansion $c = (c_1, \ldots, c_{n_b})^T$; a (n_p, n_b)-matrix Φ whose elements are $\Phi_{jk} = \phi_k(t_j)$; and a (n_b, n_b)-matrix $\mathbf{R}$ whose elements are:

$$R_{kl} = \langle D^2 \phi_k, D^2 \phi_l \rangle_{L_2(T)} = \int_T D^2 \phi_k(t) D^2 \phi_l(t) dt \tag{4}$$

Finally, the problem is solved [46] according to:

$$c = (\Phi' \Phi + \lambda R)^{-1} \Phi' z \tag{5}$$

As soon as the data are in functional form, they can be analysed to identify pollution episodes and detect outliers. The functional data allow us to identify whether different periods of time such as days, weeks or months are above the mean function and how much they are deviating. Moreover, it permits the elimination of outliers which are not real; they are due to system fails. The depth concept provides a way for ordering a set of data, contained in a euclidian space, according their proximity to the centre of the sample.

The concept of depth appeared in multivariate analyses, and was created to measure the centrality of a point among a cloud of them [50,51]. Over the years, this concept began to be introduced into functional data analysis [52]. In this field, depth represents the centrality of a certain curve x_i and the mean curve is the centre of the sample. Two depth measurements very common in the context of functional data are: Fraiman-Muniz depth (FMD) [25] and H-modal depth (HMD) [52].

Therefore, the identification of outliers with a functional approach is possible with the calculation of depths. In this case, elements that have different behavioural patterns than the rest will be taken into account. The concept of depth makes it possible to work with observations, defined in a given time interval, in the form of curves, instead of having to summarise the observations of the curve into a single value, such as the mean. This method of outlier detection is based on depth measures and centrality: an element that is far from the centre of the sample will have a low depth. Thus, the curves with the least depth are the functional outliers.

Firstly, the cumulative empirical distribution function $F_{n,t}(x_i(t))$ of the values of the curves $\{x_i(t)\}_{i=1}^n$ in a certain time $t \in [a,b]$ it is contemplated. It can be defined as

$$F_{n,t}(x_i(t)) = \frac{1}{n} \sum_{k=1}^n I(x_k(t) \leq x_i(t)) \quad (6)$$

with $I(\cdot)$ being a indicator function. Subsequently, the FMD for any curve x_i within a set of curves $x_1, \ldots, x_n$ is calculated as

$$FMD_n(x_i(t)) = \int_a^b D_n(x_i(t))dt \quad (7)$$

where $D_n(x_i(t))$ is the depth of the curve $x_i(t)$, $\forall t \in [a,b]$, expressed as

$$D_n(x_i(t)) = 1 - \left| \frac{1}{2} - F_{n,t}(x_i(t)) \right| \quad (8)$$

On the other hand, for HMD the functional mode is the element or curve most densely surrounded by the other curves in the dataset. HMD is defined as

$$HMD_n(x_i, h) = \sum_{k=1}^n K\left(\frac{\|X_i - X_k\|}{h} \right) \quad (9)$$

with a kernel function $K: R^+ \to R^+$, a bandwidth parameter h and $\|\cdot\|$ as the norm in a functional space. Among all the norms, in the most cases, it is used the norm L_2, expressed as

$$\|x_i(t) - x_j(t)\|_2 = \left(\int_a^b (x_i(t) - x_j(t))^2 dt \right)^{1/2} \quad (10)$$

In addition, also exist several options for the kernel functions $K(\cdot)$. A widely used one is the truncated Gaussian kernel, expressed as

$$K(t) = \frac{2}{\sqrt{2\pi}} \exp\left(-\frac{t^2}{2} \right), \quad t > 0 \quad (11)$$

In this paper, the depth chosen to identified outliers is the HMD. The bandwidth h is the value that leaves, below it, 15% of the data coming from the distribution of $\{\|x_i(t) - x_t(t)\|_2, \ i,j = 1, \ldots, n\}$ [15], and the cut-off C is selected, specifically, to have a 1% Type I error [50], according to

$$Pr(HMD_n(x_i(t)) < C) = 0.01, \quad i = 1, \ldots, n \quad (12)$$

The cut-off C has to be estimated because the distribution of the functional depth is not known. There are several ways to carry out this estimation; however, the bootstrapping method is the most appropriate for the purpose of this research [52,53]. The steps to follow are:

1. Extract, with replacement, a new sample of the original.
2. Estimate the study parameter through the statistic of this new sample.
3. Repeat the above steps a large number of times. This repetition is also known as Monte Carlo simulation; it relies on repetition to extract information from the data (see examples in [54,55]).
4. Obtain the empirical distribution of the statistic.

2.2.4. Functional Strengths

The functional data analysis (FDA) has many strong points, but the ones that allow the FDA to have a better performance in contamination analysis are the following: [18,19,24]:

- It is not necessary to know anything in advance about the distribution of the data.
- The analysis of the time sets as a unit. The sample analysed is structure in complete time units like days or years. Individually distributed values are not taken into account.
- Analysis of homogeneity. Outliers are defined differently; data which do not exceed the limit but which constantly have small deviations should be classified as outliers.
- Trend analysis. With these techniques, besides detect outliers, also it is possible to analyse situations where there are no outliers but small deviations from the normal data behaviour are observed.
- Complete analysis of the time spectrum. In classic studies, generally, the analyses are based on specific values measured in a determined set of points. The FDA, on the other hand, made it possible to work with the entire time spectrum of a continuous mode.

3. Results and Discussion

The results of the three methodologies presented in this document are presented below. All figures were obtained with the R-programming [56] and Python [57] software.

3.1. Classical Analysis

To perform the classical monitoring strategy on air quality the individual time series, descriptive statistics, box plots and autocorrelation analysis were calculated to determine if any of the values fell outside of the limits, and to analyse trends. The descriptive statistical parameters of the dataset are shown in Table 1:

Table 1. Summary descriptive statistics of hourly NO_2 concentrations from Blanchardstown air quality monitoring station in Dublin, Ireland. The statistical quartiles (Q1, Q2, Q3) and the interquartile range (IQR) are also displayed. Take into account that 0 $\mu g/m^3$ represents a missing or wrong value.

max	153.48 $\mu g/m^3$	Q1	9.56 $\mu g/m^3$
min	0 $\mu g/m^3$	Q2	20.61 $\mu g/m^3$
mean	28.47 $\mu g/m^3$	Q3	42.33 $\mu g/m^3$
mode	42.33 $\mu g/m^3$	IQR	32.77 $\mu g/m^3$
std	23.91 $\mu g/m^3$	var	571.79 $\mu g/m^3$
n	8760		

The descriptive statistical parameters in Table 1 show that the limit values are not exceeded. The next step of classical data analysis is present a time series of the hourly data in 2013 (Figure 1), ranging from the maximum value 153.48 $\mu g/m^3$ to the minimum value 0 $\mu g/m^3$. From here it is possible to say that in any moment, the hourly upper limit (200 $\mu g/m^3$) is not exceed and that the data have a high variability.

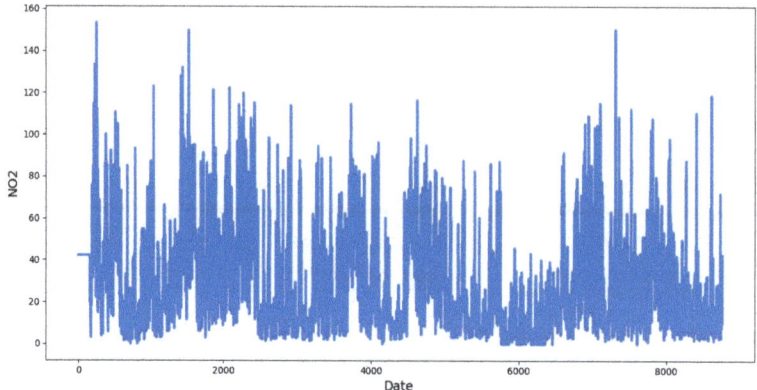

Figure 1. Individual time series of hourly NO_2 concentrations from Blanchardstown air quality monitoring station in Dublin, Ireland. Software: Python [57].

Figure 2 presents a boxplot which graphically characterises the data groups of the NO_2 concentration by quartiles. The diagram graphically displays the values of the first quartile (9.56 µg/m^3); third quartile (42.33 µg/m^3); the interquartile range (32.77 µg/m^3); and some, in red, that are considered atypical.

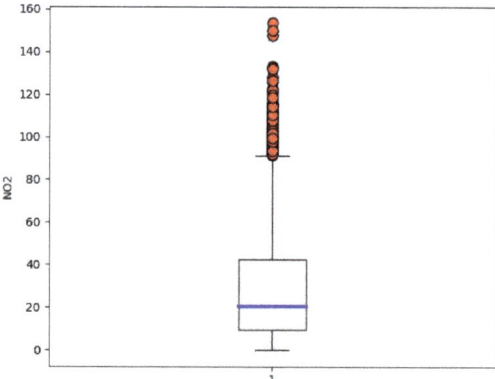

Figure 2. Box-plot of hourly NO_2 concentrations from Blanchardstown air quality monitoring station. The central and blue line represents the median, and the end of the whiskers are the quartiles (25% for the lower part and 75% for the upper part). The red dots represent the outliers. Software: Python [57].

Figure 3 presents the frequency of hourly concentrations of NO_2, which, as can be seen, are biased by 0 values. Another weakness of this analysis is that, when data are poorly collected or no data are available, only two options remain: either delete these observations (data are lost) or replace them with 0 values.

Figure 4 shows the normal probability plot of the data, again affected by 0 values. A Kolmogorov-Smirnov test and Anderson Darling test were applied to compare NO_2 concentrations to a standard normal distribution [54]. The null hypothesis is that the values have a standard normal distribution. The alternative hypothesis is that the values do not have that distribution. The results obtained for both tests were p-values very close to 0, so, with a 5% significance level, statistical evidence of the non-normality of the data has been found. The test statistic is: $\max(F(x) - G(x))$, where F(x) is the empirical cumulative distribution function and G(x) is the standard normal cumulative distribution function.

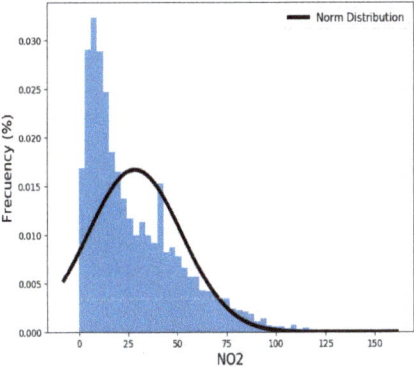

Figure 3. Frequency of hourly NO_2 concentrations from Blanchardstown air quality monitoring station. Comparison of data distribution with normal. Software: Python [57].

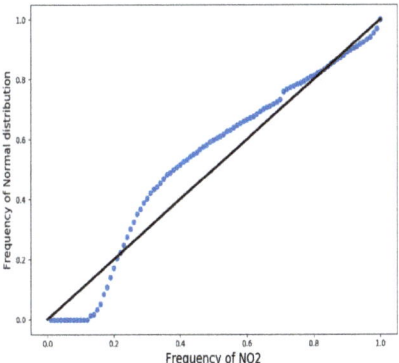

Figure 4. Normal probability plot of hourly NO_2 (QQ-plot) concentrations from Blanchardstown air quality monitoring station. Software: Python [57].

Other tests have been performed to check whether the data approaches any type of distribution: normal, generalised extreme value or Weibull and Rayleigh, but none have been acceptable with a null hypothesis at 5% significance. From the classic analysis of the data it must be concluded that there are no data that are outside the limit values. This classical method is limited to a time series analysis with regard to the assessment of trends (Figure 1), and although it allows for the identification of the main parameters within the data and how the data are distributed, is an incomplete method because it provides us with information that is too simple and does not take into account the correlation between hourly observations.

3.2. Statistical Process Control

3.2.1. Control I-MR Charts with Individual Mean

To analyse the data using the SPC method, an individual-moving range chart (IMR chart) of hourly NO_2 concentrations was made. With the examination of the results shown in Figure 5, it can be observed that the number of false alarms, i.e., outliers, is significant. This problem is attributable to:

- The non-normality of the data, which is clear from the analysis shown in Figure 4.
- The effect of autocorrelation in time series data (Figure 6).
- The existence of greater variability with data of different rational subgroups than within the data inside each analised subgroup.

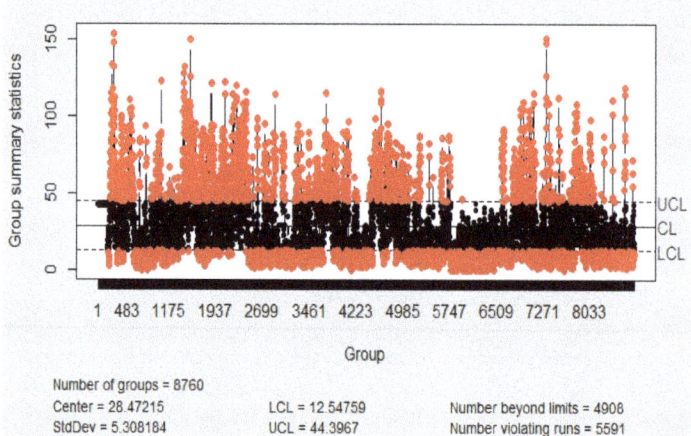

Figure 5. Individual-moving range chart (IMR chart) X/R with mobile range of hourly NO_2 concentrations from Blanchardstown air quality monitoring station. Software: R-programming [56].

By performing an autocorrelation analysis, it can be observed from Figure 6 that the data are very autocorrelated. This is very common in environmental data and shows that the autocorrelation has 24-hour cycles and decreases with time.

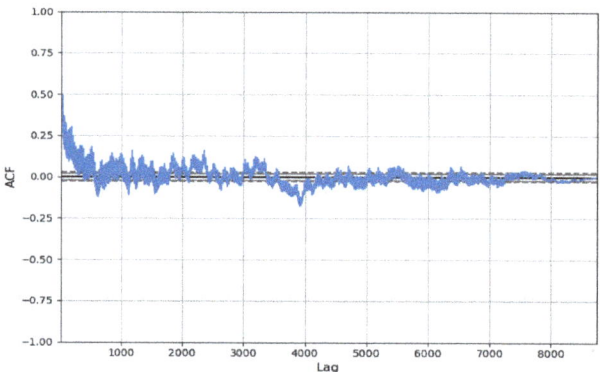

Figure 6. Sample autocorrelation function of hourly values for NO_2 concentrations from Blanchardstown air quality monitoring station. Software: Python [57].

In Figure 6 the correlation of all data for the year is shown, while Figure 7 only shows the data of the first 86 h in order to see in more detail, the 24 h cycles. Due to the non-normality of the data and the data's autocorrelation, the control chart has a large number of false alarms. Therefore, SPC is not a very suitable method with which to detect outliers for NO_2 concentration.

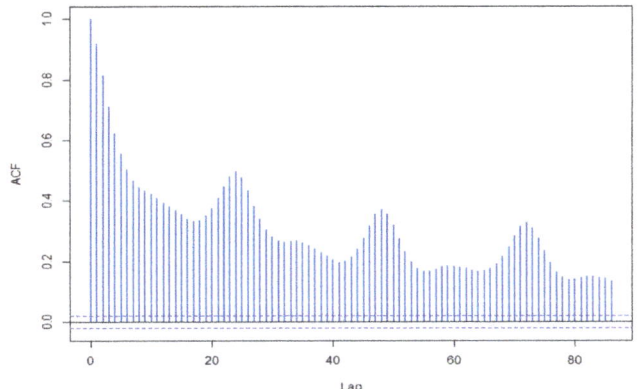

Figure 7. Sample autocorrelation function of hourly values for NO_2 concentrations from Blanchardstown air quality monitoring station over the initial 86 h period. Software: R-programming [56].

3.2.2. Control Charts with Daily Rational Subgroups

The study of datasets choosing days as the rational subgroup of the X/s chart (every day is summed up by one point), is not under control due to the non-normality and the autocorrelation (see Figures 8 and 9). Although the chart is not under control, and there is much variability, none of the 365 days exceed the limit value (100 µg/m³).

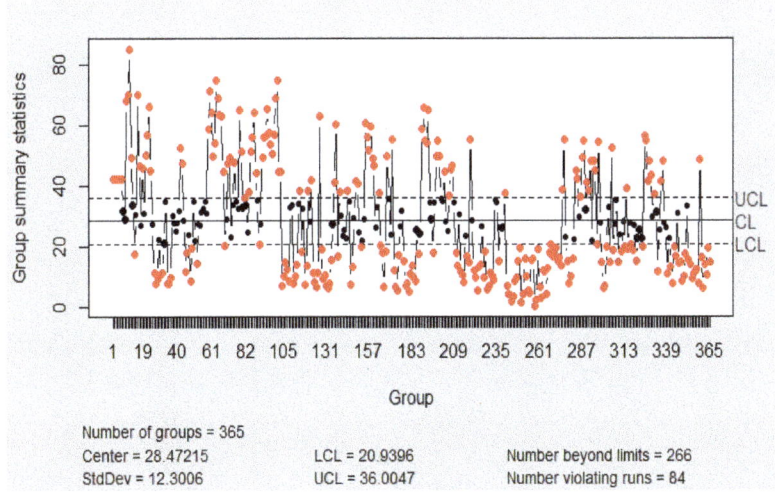

Figure 8. Xbar-chart of hourly NO_2 concentrations with the daily rational subgroup of data. Software: R-programming [56].

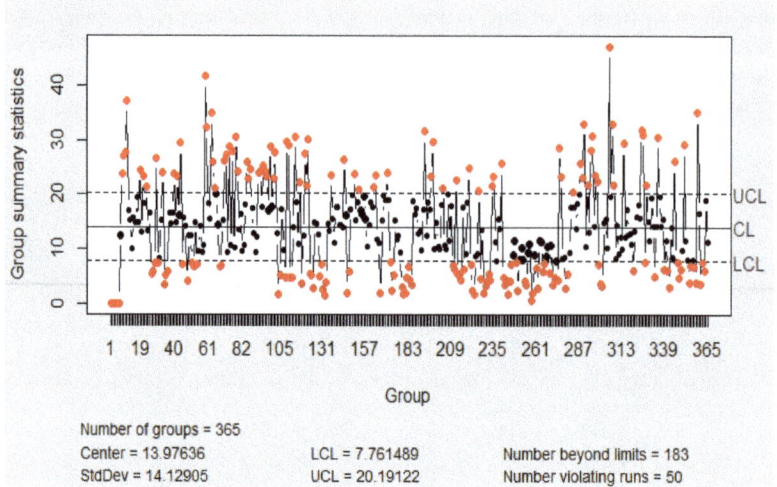

Figure 9. S-chart of hourly NO$_2$ concentrations with the daily rational subgroup of data. Software: R-programming [56].

3.2.3. Trend Analysis

Using the hourly NO$_2$ concentration data, a trend analysis was undertaken to examine the diurnal patterns and identify outliers. Figure 10 shows the box plots of hourly emissions over 365 days and represents the mean, the confidence interval of this mean, quartiles and abnormal values in red that. This way it is possible to studying individually the distribution of the NO$_2$ emissions in each hour of the day.

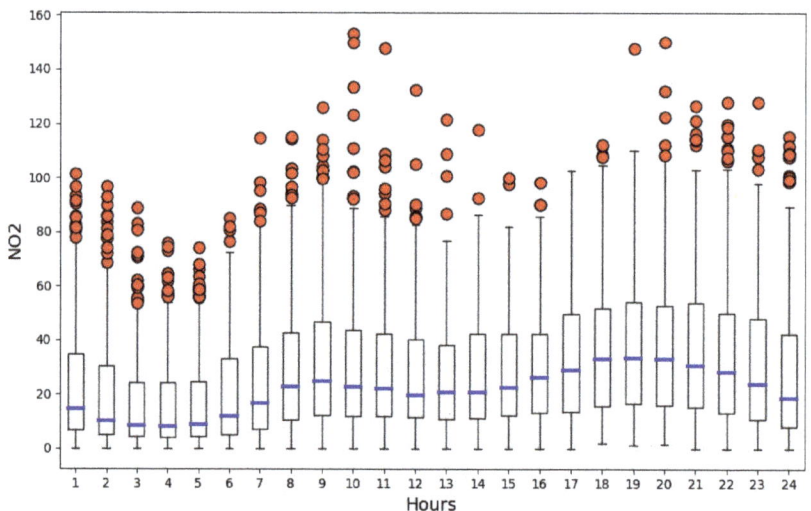

Figure 10. Hourly box plot of NO$_2$ concentrations from Blanchardstown air quality monitoring station. Software: Python [57].

In Figure 11, where only the mean values have been represented, two maximum values were analysed, at 9 a.m. and 7 p.m., which correspond to heavy traffic hours in this area of Dublin. These two hours, as can be seen in Figure 10, also have high variability (wide boxes), and around them there are numerous outliers.

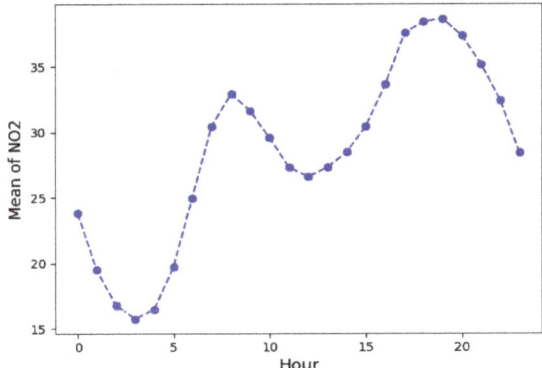

Figure 11. Mean hourly NO_2 concentrations from Blanchardstown air quality monitoring station. Software: Python [57].

It can also be seen in these figures that the absolute minimum is 4 a.m., which also corresponds to the hour with the least variability in the NO_2 concentration. If a daily analysis approach is considered throughout the year (Figure 12), it is clear that Wednesday, Thursday and Friday are the days with the highest NO_2 concentrations.

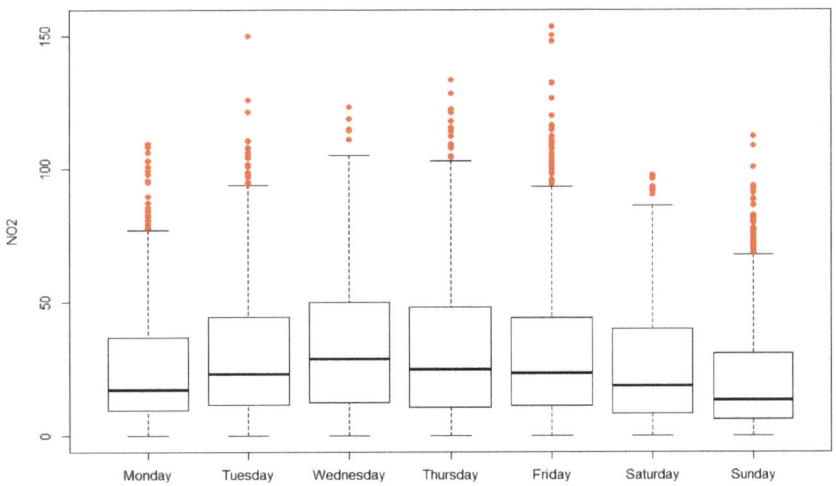

Figure 12. Daily analysis of NO_2 concentrations at Blanchardstown air quality monitoring station. Software: R-programming [56].

Blanchardstown contains the largest shopping centre in Ireland and the most major motorway nearby; because of this, there is a lot of traffic when rush hour and evening shopping are combined in this area. On the other hand, there are lower concentrations during the weekend, although not as low as expected. This is because the largest shopping centre in the country is very busy at weekends, especially during Christmas period.

In this case, a SPC gives us more information than the previous analysis, such as the differences in the NO_2 levels between days or hours. However, it still does not take into account the complete daily behaviour of NO_2 emissions from correlated hourly measurements.

3.3. Functional Analysis of NO_2 in Dublin

The subsequent step in the functional methodology is to compare the results between the classical analysis and the SPC. In the functional methodology, the first thing to do is to build a sample of curves based on the discrete values measured every hour. Figure 13 shows the 365 functions generated with 24 hourly data. Once the data are transformed into functional data, i.e., daily curves of 24 values, each of them takes into account the correlation between the hourly NO_2 values and can be analysed for outliers.

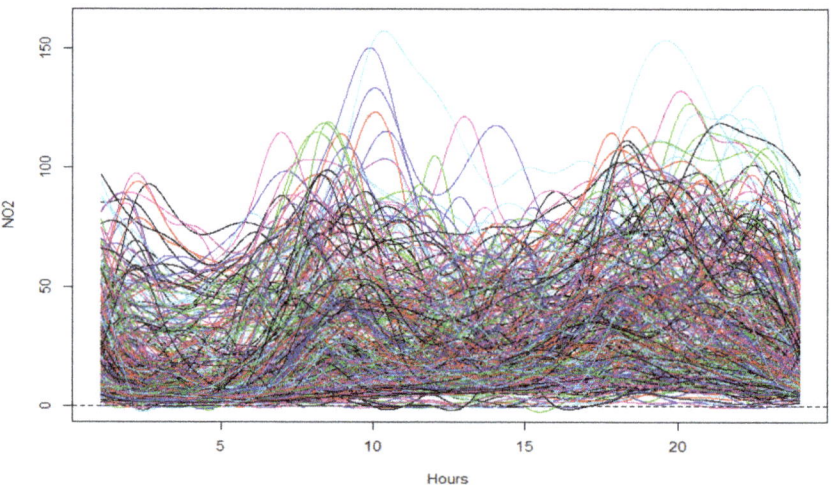

Figure 13. Data represented in functional form (functions): 365 daily curves of NO_2 emissions. Software: R-programming [56].

The results obtained with the functional analysis, taking into account the depths, allow us to identify days with abnormal functional values, even if, discreetly, they are not outliers. Despite not exceeding the daily limit values, the concentration of NO_2 over a whole day may have an abnormal behaviour. For this reason the vectorial analysis, like SPC, does not get to detect these days. In a different way, the functional approach detects any deviation from normal daily behaviour in the emissions of NO_2, without relying on any distribution restrictions. This is shown in Figure 14 where the functional outliers found in this case study are presented.

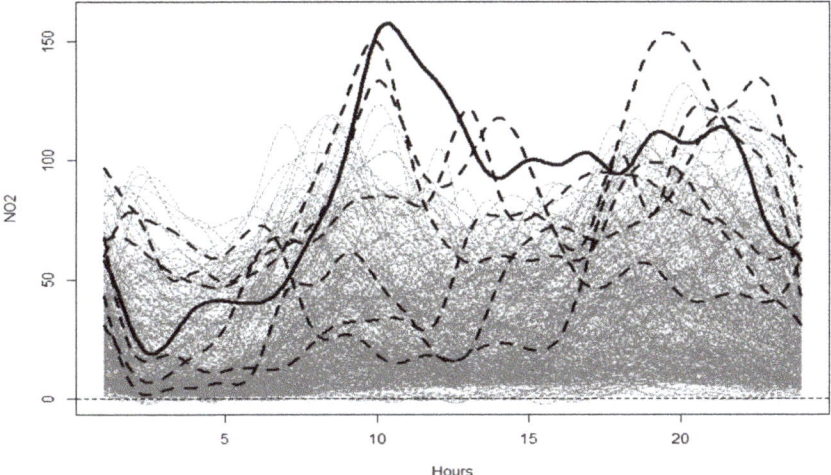

Figure 14. NO_2 functional data; and in the dark grey and dotted line, the functional outliers detected. Highlighted is the 11th day which is also an outlier. Software: R-programming [56].

The data analysed were not discretely outside the limits values; however, functionally excessive variations were observed on specific days. It can be deduced that there were no NO_2 pollution problems in 2013 because hourly, daily and annual NO_2 concentration limits were not exceeded. But it is also important to analyse whether there are hours or days with anomalous trends of NO_2 concentrations, although they remain within legal limits. The NO_x (NO, NO_2) is an pollutant not coming from a natural source ([58]), and for this reason also, its involvement in the mechanisms of depleting the ozone layer are very unfortunate. The FDA methodology has proven to be very effective in detecting days with trends that are not the same as the rest of the data. It is important not only to analyse whether the contamination is within the allowed limits, but also to find days that are different than expected.

For example, on the 11th day, detected by the functional approach and highlighted in the Figure 14, with the results obtained through the SPC, as can be seen in Figure 8, this day has a higher mean, but it is within the limit values. Neither using the classical analysis (Figure 1) with individual time series nor the one with the corresponding statistical parameters, it could be considered an outlier. Figure 14 shows a strength of the functional approach by detecting this curve as an atypical day; it will be possible to study the reasons that lead the NO_2 to behave this way on this particular day. In fact, there are several studies that demonstrate the greater power of the functional approach for detecting outliers than other methodologies (see [17,59]). There are also studies that, specifically, show that the depth measure used here (*h-modal*) is the one that achieves the lowest error rates [50].

To find a reason that explains the anomalous behaviour on those days, it would be necessary to have greater traceability of the most important sources of NO_2 emissions. It would be necessary to have data relating to the weather conditions, traffic movements, industry sources affecting the study area, etc. For example, incorporating weather conditions, i.e., temperatures, sunshine hours and precipitation, could improve the assessment of outliers. The detection of outliers and air pollution episodes can help to separate the causes of normal and specific variability, and is a first step towards the effective design and implementation of mitigation measures. Although the only reason for these outliers is possibly not the weather conditions, as can be seen in Table 2, those days were colder than usual; had very little precipitation; and in general, had fewer sunshine hours than the average for the month.

Table 2. Sample of other environmental characteristics from Dublin (temperature, sunshine and precipitation) which may impact hourly NO_2 concentrations measured at Blanchardstown air quality monitoring station. The averages shown represent the monthly average of each variable.

N°	Date	Day	Lowest T^a	Average Lowest T^a	Sunshine Hours	Average Sunshine Hours	Precipitations
10	10/01	Thursday	−2.5 °C	2.6 °C	0 h	1.2 h	2 mm
11	11/01	Friday	−0.5 °C	2.6 °C	1 h	1.2 h	0.5 mm
59	28/02	Thursday	1 °C	2 °C	0 h	2.6 h	0 mm
60	01/03	Friday	−3.1 °C	1.2 °C	3 h	2.1 h	0 mm
64	5/03	Tuesday	−3,8 °C	1.2 °C	5 h	2.1 h	0 mm
78	19/03	Tuesday	0 °C	1.2 °C	0 h	2.1 h	5.6 mm
122	02/05	Thursday	−0.4 °C	6.3 °C	5 h	6.3 h	0 mm
193	12/07	Friday	8.7 °C	11.9 °C	14 h	7.7 h	0 mm
305	01/11	Friday	2 °C	4.2 °C	4.20 h	2.7 h	0 mm

4. Conclusions

In this paper, three different analytical methods were adopted and compared to determine their effectiveness to identify pollution episodes and outliers. The data used come from a sub-urban air quality monitoring site in Dublin, Ireland, and cover the whole year 2013 with hourly measurements. Firstly, a classical vectorial approach was applied by analysing the data through time series, boxplots and frequency plots. Secondly, a statistical process control was adopted to study the data grouped by days or hours, and with different control charts (Xbar-chart, S-chart). Finally, to identify air pollution episodes and outliers, a functional data analysis approach was adopted to analyse the daily patterns of NO_2 concentrations.

To effectively support the mitigation of air pollution and provide true air quality conditions, a new approach and set of tools are required to effectively assess local air pollution. With this in mind, the classical or vectorial approach is too simplistic, despite providing significant information for decision making. It has several weaknesses related to the time correlation structure of the data, and the inability to detect real outliers, days with behaviours far from the standard, just because they do not exceed limit values. As such, more advanced and modern techniques can provide deeper insights to support controlling air pollution episodes. Statistical process control presents similar problems. Even though it manages to take into account the correlation between data, the lack of normality causes a significant increase in false alarms for days that are within the legal limits; this method labels them as outliers. This is because it works with discrete observations and it is not able to extract continuous information from the data.

Alternately, the functional data analysis method, adapted to study pollution episodes, as shown in this paper, presents important improvements as a method that can be extrapolated to any city in the world. In short, it is not restricted to certain characteristics of the data distribution; takes into account complete time units; works with the entire time spectrum of a continuous mode; and the detection of outliers is more effective, which is very important in pollution issues. On the contrary, working from a functional point of view can also have its drawbacks, such as the need for large amounts of data (it is not always possible to get them), or other data constraints, such as continuity, positivity and monotonicity. But as explained, the fact that all the hourly values of a day are not outliers does not mean that that the day is not an outlier. The outlier search with the FDA allows one to detect days that, for some reason, have different sets of measurements of NO_2 than the rest (usually higher or with strange peaks). Being able to study the reasons that make these daily functions behave differently from others makes it possible to prevent them; to try to reduce the variability in emissions; and in short, to improve air pollution control. With a classical method there would be no possibility of relating certain events to lost hours that are labelled as outliers.

The FDA's approach is presented in this document as a methodology for more effective assessment of air pollution, which is hazardous to the health of living beings, and to inform effective mitigation measures in the future.

Author Contributions: The tasks of this paper were divided as follows: on the one hand, J.P.P., J.S.V. and M.M.C. focused on the development of the mathematical methodology; on the other hand, J.M.T., A.M. and J.G. focused on the environmental part of the work. All authors have read and agreed to the published version of the manuscript.

Funding: This paper was funded by the Spanish Government (Industry and Competitiveness Ministry) under the project RTI2018-096296-B-C21.

Acknowledgments: This paper was supported by the Spanish Government (Industry and Competitiveness Ministry) under the project RTI2018-096296-B-C21.

Conflicts of Interest: The authors declare no conflict of interest.

References

1. García Nieto, P. Parametric study of selective removal of atmospheric aerosol by coagulation, condensation and gravitational settling. *Int. J. Environ. Health Res.* **2001**, *11*, 151–162. [CrossRef] [PubMed]
2. Akkoyunku, A.; Ertürk, F. Evaluation of air pollution trends in Istanbul. *Int. J. Environ. Health Res.* **2003**, *18*, 388–398.
3. Karaca, F.; Alagha, O.; Ertürk, F. Statistical characterization of atmospheric PM10 and PM2.5 concentrations at a non-impacted suburban site of Istanbul, Turkey. *Chemosphere* **2005**, *59*, 183–190. [CrossRef] [PubMed]
4. García Nieto, P. Study of the evolution of aerosol emissions from coal-fired power plants due to coagulation, condensation, and gravitational settling and health impact. *J. Environ. Manag.* **2006**, *79*, 372–382. [CrossRef] [PubMed]
5. RCPCH. *Every Breath We Take: The Lifelong Impact of Air Pollution*; Royal College of Physicians: London, UK, 2016. Available online: https://www.rcplondon.ac.uk/projects/outputs/every-breath-we-take-lifelong-impact-air-pollution (accessed on 7 February 2020).
6. WHO. *Review of Evidence on Health Aspects of Air Pollution—REVIHAAP Project*; World Health Organization: Geneva, Switzerland, 2013. Available online: http://www.euro.who.int/__data/assets/pdf_file/0004/193108/REVIHAAP-Final-technical-report.pdf (accessed on 7 February 2020).
7. Kumar, P.; Druckman, A.; Gallagher, J.; Gatersleben, B.; Allison, S.; Eisenman, T.S.; Hoang, U.; Hama, S.; Tiwari, A.; Sharma, A.; et al. The nexus between air pollution, green infrastructure and human health. *Environ. Int.* **2019**, *133*, 105181. [CrossRef]
8. EPA. United States Environmental Protection Agency. 2019. Available online: https://www.epa.gov/ (accessed on 19 December 2019).
9. AQEG. *Trends in Primary Nitrogen Dioxide in the UK*. Air Quality Expert Group. 2007. Available online: https://uk-air.defra.gov.uk/assets/documents/reports/aqeg/primary-no-trends.pdf (accessed on 7 February 2020).
10. EPA. *Ireland's Transboundary Gas Emissions*; Environmental Protection Agency: Washington, DC, USA, 2018. Available online: http://www.epa.ie/pubs/reports/air/airemissions/Irelands%20Air%20Pollutant%20Emissions%202016.pdf (accessed on 7 February 2020).
11. Costa, S.; Ferreira, J.; Silveira, C.; Costa, C.; Lopes, D.; Relvas, H.; Borrego, C.; Roebeling, P.; Miranda, A.I.; Teixeira, J.P. Integrating Health on Air Quality Assessment—Review Report on Health Risks of Two Major European Outdoor Air Pollutants: PM and NO_2. *J. Toxicol. Environ. Heal. Part B* **2014**, *17*, 307–340. [CrossRef]
12. Cooper, C.; Alley, F. *Air Pollut. Control*; Waveland Press: New York, NY, USA, 2002.
13. Lutgens, F.; Tarbuck, E. *The Atmosphere: An Introduction to Meteorology*; Prentice Hall: New York, NY, USA, 2001.
14. Jeanjean, A.; Gallagher, J.; Monks, P.; Leigh, R. Ranking current and prospective NO_2 pollution mitigation strategies: An environmental and economic modelling investigation in Oxford Street, London. *Environ. Pollut.* **2017**, *225*, 587–597. [CrossRef]
15. Cuevas, A.; Fraiman, R. A plug-in approach to support estimation. *Ann. Stat.* **1997**, *25*, 2300–2312. [CrossRef]
16. Matías, J.; Ordóñez, C.; Taboada, J.; Rivas, T. Functional support vector machines and generalized linear models for glacier geomorphology analysis. *Int. J. Comput. Math.* **2009**, *86*, 275–285. [CrossRef]

17. Martínez, J.; Garcia Nieto, P.; Alejano, L.; Reyes, A. Detection of outliers in gas emissions from urban areas using functional data analysis. *J. Hazard. Mater.* **2011**, *186*, 144–149.
18. Martínez, J.; Saavedra, A.; García Nieto, P.; Piñeiro, J.; Iglesias, C.; Taboada, J.; Sancho, J.; Pastor, J. Air quality parameters outliers detection using functional data analysis in the Langreo urban area (Northern Spain). *Appl. Math. Comput.* **2014**, *241*, 1–10. [CrossRef]
19. Sancho, J.; Iglesias, C.; Piñeiro, J.; Martínez, J.; Pastor, J.; Araújo, M.; Taboada, J. Study of water quality in a spanish river based on statistical process control and functional data analysis. *Math. Geosci.* **2016**, *48*, 163–186. [CrossRef]
20. Dombeck, D.; Graziano, M.; Tank, D. Functional clustering of neurons in motor cortex determined by cellular resolution imaging in awake behaving mice. *J. Neurosci.* **2009**, *29*, 13751–13760. [CrossRef] [PubMed]
21. Wu, D.; Huang, S.; Xin, J. Dynamic compensation for an infrared thermometer sensor using least-squares support vector regression (LSSVR) based functional link artificial neural networks (FLANN). *Meas. Sci. Technol.* **2008**, *19*, 105202.1–105202.6. [CrossRef]
22. Ordoñez, C.; Martínez, J.; Cos Juez, J.; Sánchez Lasheras, F. Comparison of GPS observations made in a forestry setting using functional data analysis. *Int. J. Comput. Math.* **2011**, *89*, 402–408. [CrossRef]
23. Ordóñez, C.; Martínez, J.; Saavedra, A.; Mourelle, A. Intercomparison Exercise for Gases Emitted by a Cement Industry in Spain: A Functional Data Approach. *J. Air Waste Manag. Assoc. (1995)* **2011**, *61*, 135–141. [CrossRef] [PubMed]
24. Sancho, J.; Pastor, J.; Martínez, J.; García, M. Evaluation of Harmonic Variability in Electrical Power Systems through Statistical Control of Quality and Functional Data Analysis. *Procedia Eng.* **2013**, *63*, 295–302. [CrossRef]
25. Fraiman, R.; Muniz, R. Trimmed means for functional data. *Test* **2001**, *10*, 419–440. [CrossRef]
26. Piñeiro, J.; Martínez, J.; García Nieto, P.; Alonso, J.; Díaz, C.; Taboada, J. Analysis and detection of outliers in water quality parameters from different automated monitoring stations in the Miño river basin (NW Spain). *Ecol. Eng.* **2013**, *60*, 60–66.
27. Sancho, J.; Martínez, J.; Pastor, J.; Taboada, J.; Piñeiro, J.; García Nieto, P. New methodology to determine air quality in urban areas based on runs rules for functional data. *Atmos. Environ.* **2014**, *83*, 185–192. [CrossRef]
28. Grubbs, F.E. Procedures for Detecting Outlying Observations in Samples. *Technometrics* **1969**, *11*, 1–21. [CrossRef]
29. Jäntschi, L. A test detecting the outliers for continuous distributions based on the cumulative distribution function of the data being tested. *Symmetry* **2019**, *11*, 835. [CrossRef]
30. EPA. *Air Quality in Ireland 2018*; Environmental Protection Agency: Washington, DC, USA, 2019. Available online: http://www.epa.ie/pubs/reports/air/quality/Air%20Quality%20In%20Ireland%202018.pdf (accessed on 7 February 2020).
31. EPA. *Air Quality in Ireland 2013: Key Indicators of Ambient Air Quality*; Environmental Protection Agency: Washington, DC, USA, 2014. Available online: https://www.epa.ie/pubs/reports/air/quality/Air%20Quality%20Report%202013.pdf (accessed on 7 February 2020).
32. Romer, U. Weather Online(Ireland). 2013. Available online: https://www.weatheronline.co.uk/weather/maps/current?TYP=tmin&KEY=IE&LANG=en&ART=tabelle&JJ=xxxx&SORT=2&INT=24 (accessed on 19 December 2019).
33. Carslaw, D.; Ropkins, K. Openair—An R package for air quality data analysis. *Environ. Model. Softw.* **2012**, *27–28*, 52–61. [CrossRef]
34. Piatesky-Shapiro, G.; Frawley, W. *Knowledge Discovery in Databases*; MIT Press: Cambridge, MA, USA, 1991.
35. Takeuchi, J.; Yamanishi, K. A unifying framework for detecting outliers and change points from time series. *IEEE Trans. Knowl. Data Eng.* **2006**, *18*, 482–492. [CrossRef]
36. Sim, C.; Gan, F.; Chang, T. Outlier Labeling With Boxplot Procedures. *J. Am. Stat. Assoc.* **2005**, *100*, 642–652. [CrossRef]
37. Montgomery, D. *Design and Analysis of Experiments*; John Wiley & Sons, Inc.: Hoboken, NJ, USA, 2013; Chapter 1–2, pp. 1–65.
38. Shewhart, W. *Economic Control of Quality of Manufactured Product*; Van Nostrand Company: New York, NY, USA, 1931.
39. Chen, Y.K. An evolutionary economic-statistical design for VSIXcontrol charts under non-normality. *J. Adv. Manuf. Technol.* **2003**, *22*, 602–610. [CrossRef]

40. Freeman, J.; Modarres, R. Inverse Box-Cox: the power-normal distribution. *Stat. Probab. Lett.* **2006**, *76*, 764–772. [CrossRef]
41. Box, G.; Cox, D. An analysis of transformations. *J. R. Stat. Soc. Ser. B (Stat. Methodol.)* **1964**, *26*, 211–252. [CrossRef]
42. Grant, E.; Leavenworth, R. *Statistical Quality Control*; McGraw-Hill: New York, NY, USA, 1998.
43. Champ, C.; Woodall, W. Exact results for Shewhart control charts with supplementary runs rules. *Technometrics* **1987**, *29*, 393–399. [CrossRef]
44. Zhang, M.; Lin, W.; Klein, S.; Bacmeister, J.; Bony, S.; Cederwall, R.; Del Genio, A.; Hack, J.; Loeb, N.; Lohmann, U.; et al. Comparing clouds and their seasonal variations in 10 atmospheric general circulation models with satellite measurements. *J. Geophys. Res.* **2005**, *110*. [CrossRef]
45. Electric, W. *Statistical Quality Control Handbook*; AT&T Technologics: Indianapolis, Indiana, 1956.
46. Ramsay, J.; Silverman, B. *Functional Data Analysis*; Springer: New York, NY, USA, 2005.
47. Sánchez-Lasheras, F.; Ordóñez, C.; Garcia Nieto, P.J.; García-Gonzalo, E. Detection of outliers in pollutant emissions from the Soto de Ribera coal-fired plant using Functional Data Analysis: A case study in northern Spain. *Proceedings* **2018**, *2*, 1473. [CrossRef]
48. Muñiz, C.D.; García Nieto, P.J.; Alonso Fernández, J.R.; Torres, J.V.; Taboada, J. Detection of outliers in water quality monitoring samples using functional data analysis in San Esteban estuary (Northern Spain). *Sci. Total. Environ.* **2012**, *439*, 54–61. [CrossRef] [PubMed]
49. Kamada, M.; Toraich, K.; Mori, R. Periodic spline orthonormal bases. *J. Approx. Theory* **1988**, *55*, 27–34. [CrossRef]
50. Febrero, M.; Galeano, P.; González-Manteiga, W. Outlier detection in functional data by depth measures, with application to identify abnormal NOx levels. *Environmetrics* **2008**, *19*, 331–345. [CrossRef]
51. Zuo, Y.; Serfling, R. General notions of statistical depth function. *Ann. Stat.* **2000**, *28*, 461–482. [CrossRef]
52. Cuevas, A.; Febrero, M.; Fraiman, R. On the use of the bootstrap for estimating functions with functional data. *Comput. Stat. Data Anal.* **2006**, *51*, 1063–1074. [CrossRef]
53. Cuevas, A. A partial overview of the theory of statistics with functional data. *J. Stat. Plan. Inference* **2014**, *147*, 1–23. [CrossRef]
54. Jäntschi, L.; Bolboacă, S.D. Computation of Probability Associated with Anderson-Darling Statistic. *Mathematics* **2018**, *6*, 88. [CrossRef]
55. Jäntschi, L.; Bolboacă, S.D. Rarefaction on natural compound extracts diversity among genus. *J. Comput. Sci.* **2014**, *5*, 363–367. [CrossRef]
56. R Core Team. *R: A Language and Environment for Statistical Computing*; R Foundation for Statistical Computing: Vienna, Austria, 2014.
57. Van Rossum, G.; Drake, F.L., Jr. *Python Reference Manual*; Centrum voor Wiskunde en Informatica: Amsterdam, The Netherlands, 1995.
58. Cosma, C.; Suciu, I.; Jäntschi, L.; Bolboaca, S. Ion-Molecule Reactions and Chemical Composition of Emanated from Herculane Spa Geothermal Sources. *Int. J. Mol. Sci.* **2008**, *9*, 1024–1033. [CrossRef]
59. Yu, G.; Zou, C.; Wang, Z. Outlier Detection in Functional Observations With Applications to Profile Monitoring. *Technometrics* **2012**, *54*, 308–318. [CrossRef]

© 2020 by the authors. Licensee MDPI, Basel, Switzerland. This article is an open access article distributed under the terms and conditions of the Creative Commons Attribution (CC BY) license (http://creativecommons.org/licenses/by/4.0/).

Article

Constructing a Control Chart Using Functional Data

Miguel Flores [1], Salvador Naya [2], Rubén Fernández-Casal [3], Sonia Zaragoza [4], Paula Raña [3] and Javier Tarrío-Saavedra [5,*]

[1] MODES Group, Department of Mathematics, Escuela Politécnica Nacional, 170517 Quito, Ecuador; miguel.flores@epn.edu.ec
[2] MODES Group, CITIC, ITMATI, Department of Mathematics, Escola Politécnica Superior, Campus Industrial, Universidade da Coruña, Mendizábal s/n, 15403 Ferrol, Spain; salva@udc.es
[3] MODES Group, CITIC, Department of Mathematics, Faculty of Computer Science, Campus de Elviña, Universidade da Coruña, 15008 A Coruña, Spain; ruben.fcasal@udc.es (R.F.-C.); paula.rana@udc.es (P.R.)
[4] PROTERM Group, Department of Naval and Industrial Engineering, Campus Industrial, Universidade da Coruña, Mendizábal s/n, 15403 Ferrol, Spain; szaragoza@udc.es
[5] MODES Group, CITIC, Department of Mathematics, Escola Politécnica Superior, Campus Industrial, Universidade da Coruña, Mendizábal s/n, 15403 Ferrol, Spain
* Correspondence: javier.tarrio@udc.es

Received: 12 October 2019; Accepted: 23 December 2019; Published: 2 January 2020

Abstract: This study proposes a control chart based on functional data to detect anomalies and estimate the normal output of industrial processes and services such as those related to the energy efficiency domain. Companies providing statistical consultancy services in the fields of energy efficiency; heating, ventilation and air conditioning (HVAC); installation and control; and big data for buildings, have been striving to solve the problem of automatic anomaly detection in buildings controlled by sensors. Given the functional nature of the critical to quality (CTQ) variables, this study proposed a new functional data analysis (FDA) control chart method based on the concept of data depth. Specifically, it developed a control methodology, including the Phase I and II control charts. It is based on the calculation of the depth of functional data, the identification of outliers by smooth bootstrap resampling and the customization of nonparametric rank control charts. A comprehensive simulation study, comprising scenarios defined with different degrees of dependence between curves, was conducted to evaluate the control procedure. The proposed statistical process control procedure was also applied to detect energy efficiency anomalies in the stores of a textile company in the Panama City. In this case, energy consumption has been defined as the CTQ variable of the HVAC system. Briefly, the proposed methodology, which combines FDA and multivariate techniques, adapts the concept of the control chart based on a specific case of functional data and thereby presents a novel alternative for controlling facilities in which the data are obtained by continuous monitoring, as is the case with a great deal of process in the framework of Industry 4.0.

Keywords: functional data analysis; statistical process control; control chart; data depth; nonparametric control chart; energy efficiency

1. Introduction

Generally, univariate and multivariate control charts are applied to identify anomalies in the industry and control the quality of products and services. However, the specific characteristics of data obtained by continuous monitoring (which has become the main trend due to advancements in sensoring and communications in the framework of Industry 4.0), require the use of increasingly sophisticated tools that take into account the presence of autocorrelation in critical to quality (CTQ) variables for the process under study.

Recently, diverse solutions have been provided that propose the modification and application of exponentially weighted moving average chart (EWMA) control charts [?]; control charts based on the fitting of autoregressive integrated moving average (ARIMA) models [?]; and the use of control charts for profiles, which is understood as the control of the parameters that define the relationship between two different CTQ variables [? ? ? ?]. Additionally, there are solutions that propose the use of techniques based on the control chart concept for anomaly detection such as machine learning techniques (neural networks and support vector machines, among others) and time series [? ? ? ? ? ? ?]. The increasingly common use of these tools is attributed to the fact that they adapt very well to the new paradigm of data (in the framework of Industry 4.0) defined by the continuous monitoring of numerous variables. Traditional control charts cannot be applied to many of the new cases due to their non-compliance with their starting hypotheses. In the domain of statistical quality control, the use of the concept of latent variables along with multivariate control charts is one of the most popular and useful alternatives to solve the problem of process control in the Industry 4.0. domain [? ? ?].

Many of these new data, usually curves, can be studied as functional data. This is the case of data in the fields of energy consumption, indoor and machine temperatures, relative humidity, amount of CO_2, among other variables, measured in all types of buildings. These new data, which currently characterize industrial processes, require innovative solutions, developed by researchers in the field of statistical quality control, based on the application of functional data analysis (FDA) techniques. Until now, few works have been published (in relation to the importance of the topic and the frequent monitoring of this type of data) on controlling the quality of processes when the data to be monitored are functional CTQ variables. Among the most outstanding research in this context, Reference [?] compared the performances of different control charts used for monitoring functional data; these charts are often identified as profiles in the statistical quality control literature. According to References [? ?], in the case of the monitoring of profiles, a set of statistical techniques (usually multivariate) is applied for controlling processes when they are defined by the functional relationship between two variables. However, a control chart for monitoring functional data, which is based on bootstrap resampling, is also proposed in Reference [?], whereas, in References [? ?], two complete monographs are proposed in which the concepts of control charts for profiles are adapted to the context of the functional data. Moreover, recently, new alternatives for outlier detection based on interlaboratory studies and those with application in industrial anomaly detection have been performed in the FDA framework [? ? ?].

This study intends to provide an alternative solution to the detection of out-of-control anomalies or states in the area of energy efficiency in buildings, specifically in commercial areas such as different stores of textile companies. Therefore, an important aim of this work is to solve the problem of such companies, specifically the ones providing lighting facilities; office automation; and heating, ventilation and air conditioning (HVAC). One such company is Fridama SL. The group comprises Fridama's facilities, Σqus (web platforms for big data management) and Nerxus (statistical consultancy for data analysis in the energy sector); the solution to the problem raised by this company is defined by continuously monitored data over time that can be treated as profiles or functional data. Based on this need, the present study proposes methodologies to build control charts that allow us to control the aforementioned processes. in order to test the applied statistical methodologies for anomaly detection properly, these companies have provided a database on a real HVAC installation, whose anomalies and assignable causes are identified by its maintenance personnel.

These energy efficiency data are non-Gaussian and autocorrelated, as the main component of continuously monitored data in the Industry 4.0 framework. Thus, the present proposal provides alternative control charts for Phases I and II of the statistical process control that can deal with non-Gaussian and autocorrelated data. As is well-known, using control charts, the process can be controlled in the following two phases: Phase I involves the stabilization or the calibration of the process and Phase II focuses on process monitoring. For Phase I, a control chart is proposed based on the depth measurement of a functional datum and the idea of atypical detection; in Phase II, a nonparametric range control chart is proposed, based on the calculation of the functional data depth, to monitor the process of interest. The procedure of building control charts corresponding to Phases I and II are explained in the next section; it also shows the results of their performance through a simulation study including diverse scenarios. The proposed methodology has been developed and programmed in R through different functions integrated in the "Quality Control Review" package, qcr, which can be freely and easily accessed by the practitioners.

Additionally, it provides a way to visualize the control charts for functional data, including the original data and the curves corresponding to the estimated control limits. This visual tool allows users and maintenance managers to relate each anomaly to an intuitively assignable cause. The results of the simulation study and its application to the real data show the usefulness of this control chart methodology in detecting anomalies when the process is defined by functional data, specifically the daily curves of energy consumption in commercial areas.

1.1. Alternatives of the Statistical Quality Control When the Basic Assumptions of the Control Charts Are Not Met

This control chart approximation for functional data is based on nonparametric multivariate control charts, which are useful when the assumptions of Gaussianity are not met. Therefore, a brief introduction of nonparametric control charts and statistical tools for dealing with autocorrelated data are presented in the following section.

Recently, analyses of the robustness of different control charts against the non-compliance of Gaussianity hypothesis have been developed. Particularly, several nonparametric control charts have been proposed. In this domain, we should highlight the works of Regina Liu [? ? ?], who has developed the r, Q and S control charts based on the data depth and rank concepts and the rank or ranges. In this regard, it is important to highlight the strong influence of Liu's works; it must be noted that several nonparametric alternatives for control charts are based on the concepts of data depth and rank. Moreover, it should be emphasized that one of the most important lines of research of the SQC, the profiles' control charts, are based, in many cases, on the application of nonparametric or semiparametric regression models [? ?]. Additionally, it is also important to highlight the use of resampling techniques for calculating natural control limits for different types of control charts [? ?]. The work in Reference [?] constitutes a complete monograph on the current trends for the development of control charts.

Despite their important advantages (no assumptions on the probability distribution of CTQ variables), nonparametric control charts are not predominantly used in the industrial and business framework. As noted in Reference [?], this can be attributed to several factors such as the lack of specific software, both commercial and free; the lack of general training in nonparametric statistics that generates insecurity and distrust in users and the lack of contrasted reference texts for the application of nonparametric methods in SQC. However, the research activity in this field is growing, as illustrated by the work in Reference [?].

Another starting hypothesis of control charts is the independence of the observations. The data continuously monitored over time by different sensors usually show a variable level of autocorrelation (the greater the correlation the closer the observations are in time). The application of standard techniques in the case of the violation of the independence hypothesis often results in the detection of an unacceptable number of false alarms [?]. Therefore, the development and analysis of techniques that remove the sample autocorrelation is fully justified. Within these techniques, the most widespread is the application of time series models (e.g., autoregressive moving average (ARMA) and ARIMA) to remove the correlation between observations and the subsequent monitoring of the error variable (difference between actual values and those estimated by the model) using control charts [? ? ?]. Moreover, References [? ?] propose the combination of control charts with adjustment algorithms. Finally, the monographs of [? ?] describe the most relevant lines of research for controlling and monitoring autocorrelated data.

It is important to note that this type of data is related to the control charts for functional data. The use of FDA techniques allows us to consider the autocorrelation of the data as well as, by means of resampling techniques, to circumvent parametric assumptions about the trend and dependence.

1.2. Control Charts for Phases I and II

The studied process is stabilized in the framework of the Phase I control charts, that is, the process is left under control [?]. This implies that there are no other assignable causes in the process, except for those present due to randomness itself (non assignable causes). This is equivalent to stating that the process remains stable—the parameters of the probability distribution of the CTQ characteristic remain unchanged. This allows us to estimate the natural control limits for the variable that describes the quality of the process. The natural control limits are estimated using a preliminary or calibration sample. Thus, in this stage, the main assignable causes of variation are identified and rectified through corrective actions. After the variations are removed, we estimate the natural control limits of the CTQ variable corresponding to the process under control. While most of the literature in the field has referred to Phase II control charts, within the new data paradigm for Industry 4.0, it would be essential to develop methodologies for Phase I [?]. In this regard, Reference [?] presents an overview of the recent contributions to the development of Phase I control charts. Among the latest proposals for control charts, the work of Grasso et al. [?] is especially interesting taking into account that is a Phase I control chart proposal for profiles based on the use of functional data depth concept (as well as the case of the present proposal). In fact, they proposed a Phase I control chart methodology for profiles belonging to the "multi-modelling framework", that includes the following stages: (1) a classification stage of new profiles into different operating modes or profile patterns using functional classification techniques from functional data depth measures (maximum depth approach based on Mode depth); (2) even the identification of a novel operation mode is included; (3) Once the operation mode corresponding to a new profile is identified, this is assigned to the corresponding control chart and consequently a suitable control charting method is applied to determine if the process remained in control over the period of time where those data were collected.

In this study, the atypical detection procedure presented in Reference [?] is adapted to develop a control chart for Phase I; this approach allows us to obtain a calibration sample from an under-control process that can be monitored through new measurements using a Phase II rank control chart. In fact, in Phase II, it is assumed that the process is under control [?]; additionally, in each new sample (monitored sample), the statistical rank is obtained and it is represented in the control chart with the estimated lower control limit.

2. Methodology

In this section, methodologies of control charts for functional random variables, $\mathbf{X}$, which take values in a functional space $\mathcal{E} = L^2(T)$, with $T \subset \mathbb{R}$, are developed.

Based on observations of the functional variable $\mathbf{X}$, we obtain a sample each of calibration and monitoring; they are functional datasets of sizes n and m, respectively, which allow us to build control charts for Phase I (in the case of the calibration sample) and Phase II (from the monitoring sample).

In the case of designing the control charts, any unstable or out-of-control process is refereed to the assignable causes of variation emerging from unusual and avoidable events that interrupt the process, that is, when they cause a change in the parameters of the underlying model of the profile or functional data [?]; these variations can be eliminated from the data by identifying and acting on the cause; this approach will avoid such variations in the future [?].

Concerning a method for building quality control charts, the probability of process instability (level of significance) presents a measure of its performance. This probability, provided that it is within the H_0, allows the derivation of at least one measure (observed value of the statistic) outside the control limits [?]. Phase I involves the development of a method and the estimation of the level of significance;

however, in the case of Phase II, is the process is assumed to be under control, that is, the level of significance is fixed.

2.1. Procedure for Building a Control Chart for Phase I (Stabilization)

As mentioned above, in Phase I, the retrospective data corresponding to a calibration sample of size n is analyzed for evaluating the stability of the process whose quality is ascertained over time and for estimating the parameters of the control chart [?].

In Phase I, a control chart is used to test the hypothesis that there is no change in the distribution of observations of the variable ordered with respect to time $\{\mathcal{X}_1(t), \mathcal{X}_2(t), \ldots, \mathcal{X}_n(t)\}$.

These changes can be punctual (freaks or bunches) or they can be related to a change in process that is evaluated (observable through patterns of sudden or gradual change in the mean of the process). Concerning isolated changes, it refers to the occurrence of at least one observation of the observed variable that deviates from the distribution of the other observations [?]. The hypothesis tested in Phase I is:

$$H_0 : \mathcal{X}_i(t) \stackrel{d}{=} \mathcal{X}_j(t), \forall i, j \in \{1, \ldots, n\}$$
$$H_a : \mathcal{X}_i(t) \stackrel{d}{\neq} \mathcal{X}_j(t), \text{for some } i, j \in \{1, \ldots, n\}. \quad (1)$$

The stabilization phase of a process consists of applying an iterative method that allows the detection and elimination of those observations (in this context curves) that have a deviation with respect to the shape or magnitude of most of the observed curves. In other words, a curve is an atypical value if it has been generated by a different stochastic process or there is a change in the trend or variability of the stochastic process with respect to that corresponding to the remaining data [?]. The quantity of outliers is assumed to be unknown, although small.

In this study, the proposed control charts for Phase I are based on the adaptation of outlier detection methodologies from functional data, based on the data depth calculations [?].

The method of the detection of outliers for functional data [?] considers an atypical curve if its depth is less than a specific quantile of the distribution of depths estimated by bootstrap. In other words, an atypical curve will have a significantly low depth.

This procedure can be used with different types of functional depths. In the library fda.usc [?], the following alternatives are offered: Fraiman and Muniz (FM) [? ?], mode depth [?] and random projections depth (RP) [? ?]. In this work, the outlier detection procedure proposed by Febrero et al. fda.usc has been adapted to estimate a specific quantile of the depth distribution that plays the role of the lower control limit (LCL) for a Phase I control chart.

The control chart proposed for Phase I is estimated and plotted from a depth measurement (FM, RP or mode) and only the lower control limit (LCL) is considered to detect if the process is out-of-control (the depth of a curve is less than the LCL). In addition to this representation, an additional chart showing the original curves is proposed to provide an intuitive idea about the cause behind the identified anomaly (by checking its shape or magnitude) and thus to identify assignable causes. For instance, in the case of the HVAC energy consumption in buildings, anomalies may include stopping air handler, failure of the counter or sensor, a change in the regulation of the machines and adverse weather conditions.

In Phase I, we consider the functional random variable $\mathcal{X}$, from which a random sample is drawn—$\{\mathcal{X}_1(t), \mathcal{X}_2(t), \ldots, \mathcal{X}_n(t)\}$. These data are used to formulate the following steps to build the control chart for Phase I:

1. Calculate the depth corresponding to each observation of the dataset, $D(\mathcal{X}_i)_{i=1}^n$ and make a control chart based on the depth of each datum with respect to time.
2. Choose the parameter LCL according to the significance level of the control chart, that is, the percentage of false alarms (observations under control but erroneously detected as

out-of-control) is small (for example, $\alpha = 1\%$). The following procedures are used to estimate the LCL.

- Bootstrap procedure based on Trimmed:
    - Reorder the curves according to their depths in a decreasing way. $\mathcal{X}_{(1)}, \ldots, \mathcal{X}_{(n)}$.
    - It is assumed that, at most, $\alpha\%$ of the sample can be considered outliers.
    - B samples are obtained by a smoothed bootstrap procedure from the dataset resulting from discarding the $\alpha\%$ of the less depth curves. Let $\mathcal{X}_i^{*b}, i = 1, \ldots, n, b = 1, \ldots, B$ these bootstrap samples. To obtain each bootstrap resample:
        * A uniform sampling is done, i^* of $1, \ldots, [n(1-\alpha)]$.
        * Z_{i*} is generated as a Gaussian process with zero mean and variance and covariance matrix.

          $\gamma \Sigma_\mathcal{X}$ with $\gamma \in [0,1]$, where $\Sigma_\mathcal{X}$ is the variance and covariance matrix of the observations $\mathcal{X}_{(1)}, \ldots, \mathcal{X}_{([n(1-\alpha)])}$.
        * Finally, you get $\mathcal{X}_i^{*b} = \mathcal{X}_{(i*)} + Z_{i*}$.
    - For each $b = 1, \ldots, B$, we obtain C^b, which is the $\alpha\%$ empirical quantile corresponding to the distribution of the depths, $D(\mathcal{X}_i^{*b})$. The output value, C = LCL, is the median of the values $C^b, b = 1, \ldots, B$.

- Bootstrap procedure based on weighing:
    - The depths of the $\mathcal{X}_1, \ldots, \mathcal{X}_n$ curves are obtained.
    - B samples are obtained through a smoothed bootstrap from the original dataset weighted by their depths. Let $\mathcal{X}_i^{*b}, i = 1, \ldots, n, b = 1, \ldots, B$ these bootstrap samples. These replicas would be obtained:
        * Weighted sampling is performed, with i^* of $1, \ldots, n$ and with probability proportional to $D(\mathcal{X}_1), \ldots, D(\mathcal{X}_n)$.
        * Z_{i*} is generated as a Gaussian process with zero mean and variance-covariance matrix $\gamma \Sigma_\mathcal{X}$, with $\gamma [0,1]$, where $\Sigma_\mathcal{X}$ is the variance and covariance matrix of the observations $\mathcal{X}_1, \ldots, \mathcal{X}_n$.
        * Finally, we get $\mathcal{X}_i^{*b} = \mathcal{X}_{i*} + Z_{i*}$
    - For each $b = 1, \ldots, B$, we get C^b, which is the empirical quantile corresponding to the $\alpha\%$ of the distribution of the depths, $D(\mathcal{X}_i^{*b})$. The final value C = LCL is the quantile β of the values C^b, with $b = 1, \ldots, B$.

3. If there is any curve such that $D(\mathcal{X}_i) \leq$ LCL for a given LCL, then it would be considered atypical and the process would be out-of-control.
4. Additionally, a control chart including the original curves and the functional envelope obtained from 99% of the deeper bootstrap replicas is also developed.

Moreover, once the atypical curves are detected, they are removed and the procedure is repeated until the process becomes stable (under control), namely, defined by a total absence of atypical data.

2.2. Procedure for Building a Control Chart for Process Monitoring (Phase II)

Phase II deals with process monitoring; it involves quick detection of changes from the calibrated sample stabilized in Phase I [?]. For the scalar and multivariate cases, the process is monitored by taking the estimated control limits in Phase I [?] as a reference. In this phase, the average run length (ARL) is used to evaluate the performance of the control charts [?].

In this context, we test if there are deviations between the data obtained in Phase II, also called monitoring sample, $\{\mathcal{X}_{n+1}(t), \mathcal{X}_{n+2}(t), \ldots, \mathcal{X}_m(t)\}$ and the reference data $\{\mathcal{X}_1(t), \mathcal{X}_2(t), \ldots, \mathcal{X}_n(t)\}$ or calibration sample, taking into account its distribution.

In Phase II, in the univariate case, an F distribution for the under-control process is estimated from a calibration sample or reference data. It is assumed that F is the distribution of the CTQ variable of an under-control process (Phase I). This distribution is used to establish control limits that will be used to monitor the process in Phase II. The limits comprise an interval that will cover new observations of the process with a high probability, assuming that the process is under control. In Phase II, a sample of the G distribution is monitored. Therefore, in this stage, the methods for constructing control charts are based on contrasting the hypothesis:

$$\begin{aligned} H_0: & \quad F = G \\ H_1: & \quad F \neq G. \end{aligned} \qquad (2)$$

In the FDA context, we do not have a density function for a functional random variable $\mathcal{X}$ that allows us to perform different tests corresponding to Phases I and II. Alternatively, we can estimate the distribution of the depth corresponding to the curves that belong to a sample of functional data. Thus, for Phase II, the use of the rank control charts [?] is proposed in an FDA context. The calculation of depths for functional data is proposed, which facilitate the calculation of ranks. These form the basis for developing r control charts, called rank charts.

The adaptation of r control charts involves the calculation of the rank statistic from functional depth measurements. The r chart plots the rank statistic as a function of time. The central control line CL = 0.5 serves as a reference point for observing possible patterns or trends. The lower limit is LCL = α, where α is the false alarm rate.

For the practical application of this proposal, the qcr [?] and fda.usc [?] R packages are used. The fda.usc package provides tools for the calculation of functional data depth, while the rank control chart, among other nonparametric charts proposed in Reference [?], is applied using the qcr package, which was developed by the authors.

As mentioned above, in Phase II, the curves corresponding to the calibration sample of Phase I, $\{\mathcal{X}_1(t), \mathcal{X}_2(t), \ldots, \mathcal{X}_n(t)\}$, are used for detecting changes or deviations with respect to the behavior of the process described in Phase I. The curves of the monitoring sample, $\{\mathcal{X}_{n+1}(t), \mathcal{X}_{n+2}(t), \ldots, \mathcal{X}_{n+m}(t)\}$, are collected; additionally, we test the hypothesis of each new curve belonging to the same distribution that corresponds to the calibration sample.

The procedure for estimating control charts for Phase II follows the same scheme presented in Reference [?]. Particularly, we assume that the rank statistic follows a uniform asymptotic distribution. This result is applicable to the functional case because of the way the rank corresponding to each observation is calculated (percentage of less deep curves than the observed ones). This fact provides a computational advantage in the monitoring of continuous processes since it eliminates the need to estimate the LCL. However, it is set as the quantile of a uniform distribution at a significance level α.

The procedure to develop the rank control chart for the functional univariate case (process defined by one functional variable) is detailed below, which can be easily generalized to the functional multivariate case (process defined by more than one functional variable).

1. From the reference sample $\{\mathcal{X}_1(t), \mathcal{X}_2(t), \ldots, \mathcal{X}_n(t)\}$, get the depths of the dataset, $D(\mathcal{X}_i)_{i=1}^n$ and the depths of the curves that make up the monitoring sample, $D(\mathcal{X}_j)_{j=n+1}^{n+m}$. The depth of each curve corresponding to the monitoring sample is calculated from the n curves of the calibration sample, that is, with respect to $n+1$ cases.
2. The rank statistic is calculated for each curve of the monitoring sample, $r_G(\mathcal{X}_{n+1}), \ldots, r_G(\mathcal{X}_{m+n})$, considering the calibration sample $\{\mathcal{X}_1(t), \mathcal{X}_2(t), \ldots, \mathcal{X}_n(t)\}$ as sample of reference.

$$r_G(\mathcal{X}) = \frac{\#\{\mathcal{X}_i | D(\mathcal{X}_i) \leq D(\mathcal{X}), i = 1, \ldots, n\}}{n}.$$

3. The values of the rank statistic, the lower control limit LCL = α and central line CL = 0.5 (the expected value of the rank statistic) are plotted, thereby generating the control chart.
4. Proceed to monitor the process. If at least the rank of a curve, $\mathcal{X}$, is such that $r_G(\mathcal{X}) \leq$ LCL, then the process is considered out-of-control.
5. A functional control chart is developed. This is a graphical tool that allows us to identify the possible assignable causes of the out-of-control states. The original curves are included, those correspond to the reference and monitoring samples, in addition to the functional envelope obtained from $(1 - \alpha)\%$ of the deepest curves of the calibration sample.

3. Data Collection: Case Study of HVAC Installations in Commercial Areas

Here, the case study of HVAC installations' control are considered for a clothing store of a commercial area in the Panama City [?]. The data stream has been obtained by using the Σqus energy web platform. Sixteen CTQ variables are measured taking into account their ability to provide information about the energy efficiency, air quality and the thermal comfort of the store environment—indoor temperatures, overall energy consumption, HVAC energy consumption, CO_2 content in the air (ppm), relative humidity (%), temperatures of impulsion and return temperatures of the chillers in different areas of the store (see Figure ??).

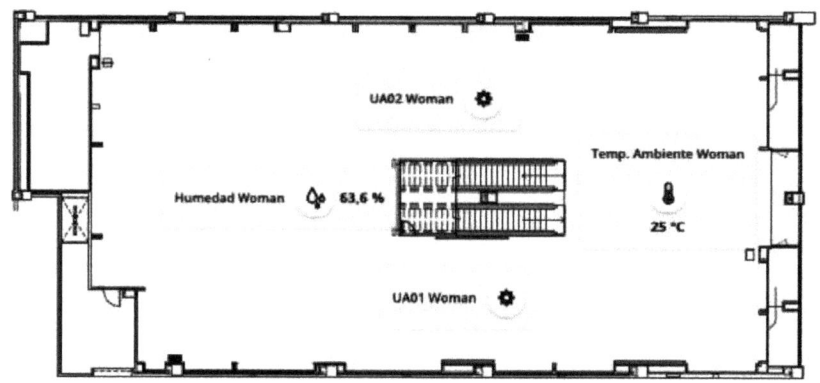

Figure 1. Plan of the case study store located in the Panama City.

Hourly measurements are obtained from 1 August 2017 to 31 October 2018. The operations of the HVAC facilities of the store start at 9:00 a.m. or 10:00 a.m. At start-up, the energy consumption peaks due to the characteristics of the HVAC installation. From 12:00 p.m., the consumption remains relatively constant the store closure at 20:00 p.m., 21:00 p.m. or 22:00 p.m. The shutdown takes about 1 or 2 h, with consumption falling at a constant rate of change. The resulting data can be considered functional data and thus FDA techniques can be applied. It is also important to note that this case study is a controlled study in which the anomalies and their assignable causes have been previously detected for the maintenance staff.

The data were obtained in the framework of a controlled environment where anomalies were identified by the maintenance staff. They are briefly described as follows:

- On 11 September, there was a decline in the air conditioning consumption at about midday.
- On 21, 22 and 30 September, the shopping center was closed and thus there was no energy consumption and the temperatures remained high.
- On 27 September, several maintenance tests were applied to the store's HVAC installations.
- On 29 September, the store's HVAC installations were stopped one hour earlier.

- Additionally, on 19 September, the air conditioning was stopped half an hour before the usual time. Particularly, there was a regulation change in the HVAC system.
- In mid-October, there was a leak in the air conditioning circuit. From that moment, energy consumption began to rise.
- On 1 November, repairing activities were performed. Consequently, the consumption decreased and the start-up consumption peak was removed.
- Between 17 and 20 November, the energy consumption in HVAC was again increased.

It is important to note that only working days have been studied in this work to evaluate the performance of the proposed control charts.

4. Application to Real Data

This section shows the usefulness and performance of the new graphical methodology for quality control using functional data, which is evaluated in the case study on the detection of energy efficiency anomalies of an HVAC installation. Specifically, the case study considers a commercial area of a well-known Galician apparel brand located in the Panama City. In this controlled case study, the anomalies and their assignable causes were detected by the maintenance personnel.

The following section shows the need to develop and apply FDA methodologies for control charts, considering the observations of the data of the present case study, particularly those corresponding to August. As mentioned above, in August, no event destabilized the process due to assignable causes. However, by using a methodology for scalar data (ignoring the autocorrelation between the variables), an unacceptable number of false alarms could be detected.

In the scalar case, boxplot [?] is commonly used to detect anomalous or atypical data. Figure ?? shows a traditional scalar approach for detecting outliers using boxplot. The left panel shows the boxplot for each variable of energy consumption in HVAC systems per hour, while the right panel shows the curves of daily energy consumption in HVAC systems, highlighting curves detected as outliers by the descriptive procedure based on the application of boxplots to each hourly consumption. In the usual procedure, atypical curves are those in which at least one point has been detected as an outlier in some boxplot; however, the drawback of this approach is that it increases the probability of type I error. It detects 12 daily energy consumption curves as outliers.

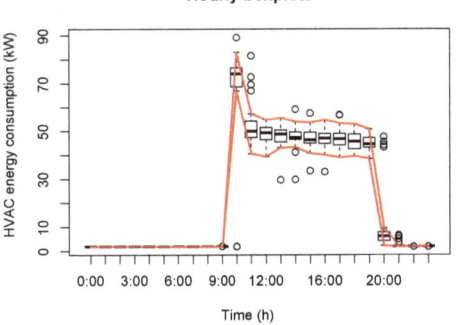
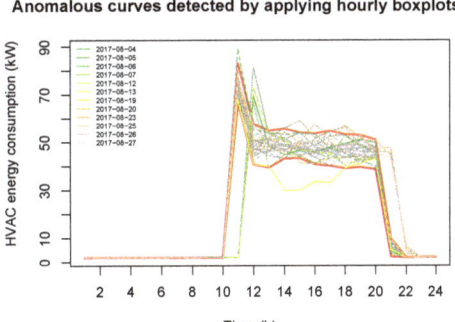

Figure 2. Detecting outliers in the heating, ventilation and air conditioning (HVAC) energy consumption by developing a boxplot for energy consumption per h.

Based on the information described in the previous section, first, we apply the data depth control chart for Phase I and, subsequently, the rank control chart to monitor the process during Phase II. The application of these two statistical techniques, together with the contribution of an intuitive graphic tool (to facilitate the detection of assignable causes for the anomalies), constitutes the new proposed procedure of control charts for functional data. Generally, the procedure can be summarized as follows:

- The energy consumption curves in HVAC during August and September accounts for the calibration sample. A broad temporal range has been taken to estimate the natural control limits of energy consumption in HVAC accurately.
- For Phase I, the reference or calibration sample is stabilized by using the control chart developed from the depth measures of the curves. Particularly, curves detected as atypical by FDA outlier data detection methods are removed from the calculations to estimate the natural control limits.
- Data corresponding to October and November are monitored to confirm that there is no deviation in the HVAC processes. If there is a change, that is, there are days with out-of-control HVAC energy consumption, then it was recommended that the possible assignable causes should be sought, which, when detected, should be removed in order to remove the corresponding process variations.

In Figure ??, the black curves correspond to August (23 curves), whereas the gray curves account for the HVAC energy consumption in September (21 curves). The days from Monday to Friday are used in this study, taking into account that the work schedule is different on Saturday and Sunday. In September, the actual anomalies were detected (see Figure ??, curves in red) using the Phase I control chart based on the functional data depth outlier detection methods. Particularly, the anomalies corresponding to 11, 21, 22, 27 and 29 September were identified (refer to Section ?? for more information on the assignable causes of these anomalies).

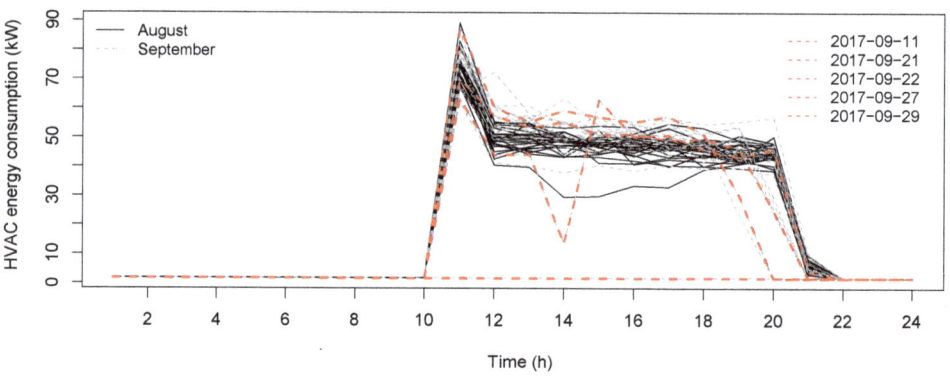

Figure 3. Daily curves of energy consumption in HVAC facilities of the Panama City's store. The natural control limits are estimated from the curves belonging to the calibration sample, which have been shown which (Phase I).

From this reference sample and based on the simulation study corresponding to the control chart based on the functional depths (see Section ??), the mode functional depth and the weighed method for outlier detection are used. Additionally, a significance level of $\alpha = 0.025$ is used for the estimation of the LCL from $B = 500$ bootstrap resamples, a smoothing coefficient $\gamma = 0.8$, and a percentage of Trimming $trim = 0.05$, which allows obtaining an envelope of 95% of the deepest curves. The advantage of this procedure of detecting atypical curves in the Phase I control chart based on functional data is its flexibility to adapt to a wide variety of real problems, by regulating its parameters (γ and $trim$).

The left panel of the Figure ?? shows the original gray curves, the estimate of the median (blue curve), the functional trimmed mean (red curve) and the envelope corresponding to the 95% of the deepest curves, which is plotted in red. The curves detected as anomalous are shown in gray and dotted in black. After the mentioned atypical energy consumption curves are identified, they are removed from any calculation related to the estimation of the natural control limits, given that their assignable causes have been identified (causes apart from the randomness of the data). Subsequently, the process is repeated according to an iterative scheme until atypical curves associated with assignable causes are not identified.

Figure ?? shows the functional approximation of the rank control chart, where each datum or point accounts for a daily curve of energy consumption in HVAC. The first 23 points represent the HVAC energy consumption curves for August, whereas the next 21 points represent the energy consumption curves of September.

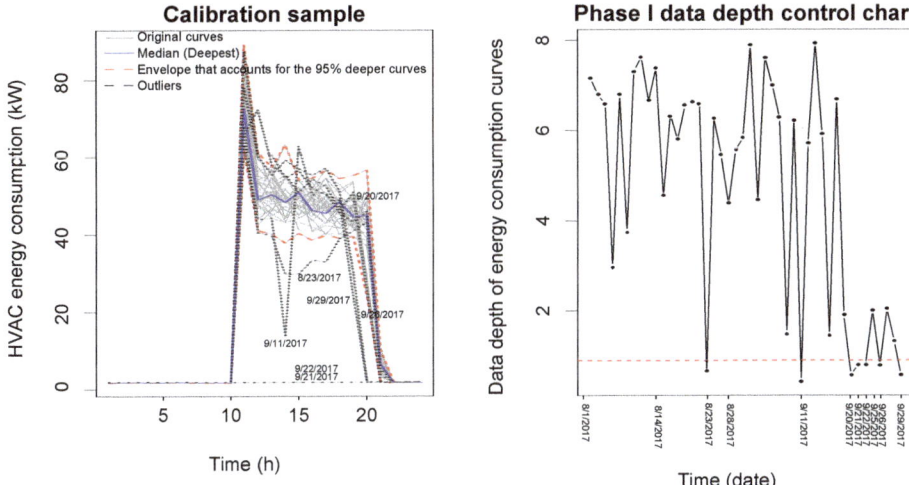

Figure 4. **Left** panel: Curves of daily energy consumption in HVAC facilities of the Panama City's store. The figure identifies curves detected as anomalies as well as functional position measurements. **Right** panel: Control chart corresponds to Phase I; they facilitate the detection of anomalous curves, thus eliminating the assignable causes of variation. The depths of each of the daily consumption curves and the natural control limit are shown.

In this first iteration, the previously identified curves are detected as anomalous. Their structure corresponds to an assignable cause of variation (see Section ??). They correspond to 11, 21, 22 and 29 September. However, in this iteration, the anomalous curve corresponding to 27 September has been not detected. Additionally, the curves corresponding to 23 August, 20 and 26 September are detected. Moreover, the data depth corresponding to the curves from 19 September are very small—they are outside the lower control limit and thus they are identified as out-of-control states. The assignable cause corresponding to this behavior is that the air conditioning was being shut off half an hour earlier than usual.

In the next iteration, the following daily consumption curves are detected as out-of-control or anomalous: 4 August, 15, 19, 25 and 27 September. In this iteration, the HVAC energy consumption curve of 27 September and those corresponding to the last days of September are identified; the timing of HVAC shutdown was changed for the latter part of September. Finally, energy consumption curves are not detected as anomalous after the second iteration (see Figure ??).

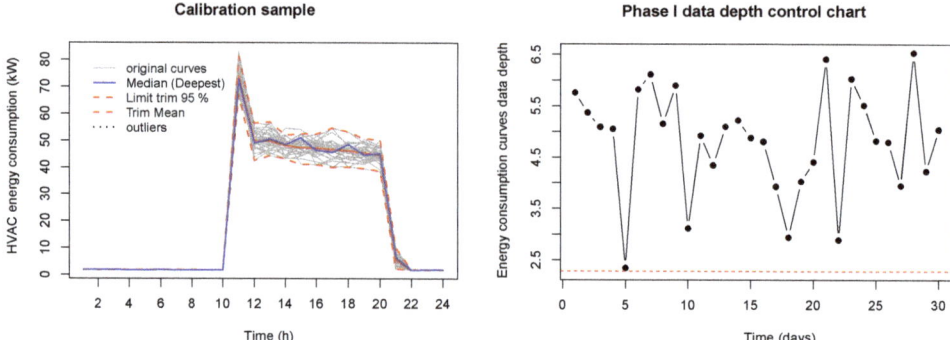

Figure 5. Under-control process (HVAC energy consumption is stabilized). **Left** panel: daily HVAC energy consumption curves in the Panama City store, showing the envelope corresponding to 95% of deeper curves. **Right** panel: control chart for Phase I based on functional data depth.

The process began with 44 curves, of which 9 (in the first iteration) and 5 (in the second iteration) curves were identified as anomalous. Of these 14 energy consumption curves, 2 and 12 curves correspond to August and September, respectively. Hence, the calibration sample comprises days in August and September, until 18 September. The current performance of energy efficiency of HVAC installation has been characterized through this reference or calibration sample. The next step is to detect changes (from the reference HVAC performance) in the HVAC system. This task belongs to Phase II control charts.

After Phase I, the Phase II control charts based on functional data, also called monitoring phase, is performed. Therefore, the sample to be monitored comprises days corresponding to October in which the HVAC facilities of the apparel store were operated at regular times. This is in consideration of the fact that a leak in the air conditioning circuit was recorded in mid-October, especially after the HVAC consumption began to rise. This behavior can be observed in Figure ??; it plots the monitored sample as well as the curves of the calibration sample and their 95% envelope, which is estimated in Phase I.

Figure ?? shows that the HVAC energy consumption, the CTQ variable for the energy efficiency of HVAC installations, is out of control. The assignable cause is the leak in the air conditioning system. In order to rectify the anomalous operation of the HVAC installations, on 1 November, a provisional repair was performed. These actions produced a decrease in the HVAC energy consumption and attenuated peak energy consumption corresponding to the start-up (see the energy consumption curves before 11:00). Consequently, the consumption of the HVAC facilities in November is slightly different from that of August and September, which account for the calibration sample. However, this energy consumption resumes its rise between 17 and 20 October.

These changes can be observed in Figure ?? wherein it is observed that the HVAC energy consumption corresponding to the monitored sample is characterized by a greater variability than that corresponding to the calibration or reference sample. The results of the application of the rank control chart show that the process is out-of-control both for the monitoring samples of October and November. In October, the assignable cause is the fault in HVAC facilities, whereas the assignable cause in November is the change in the process due to repair activities that deals on a decreasing in HVAC energy consumption.

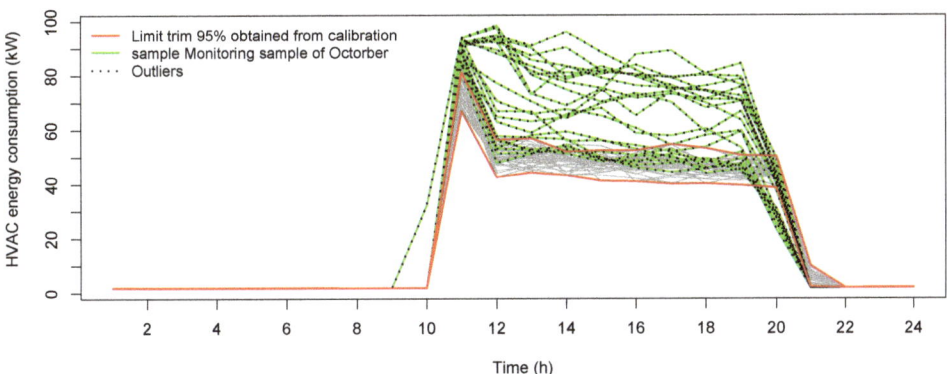

Figure 6. Reference of calibration and monitored samples corresponding to October.

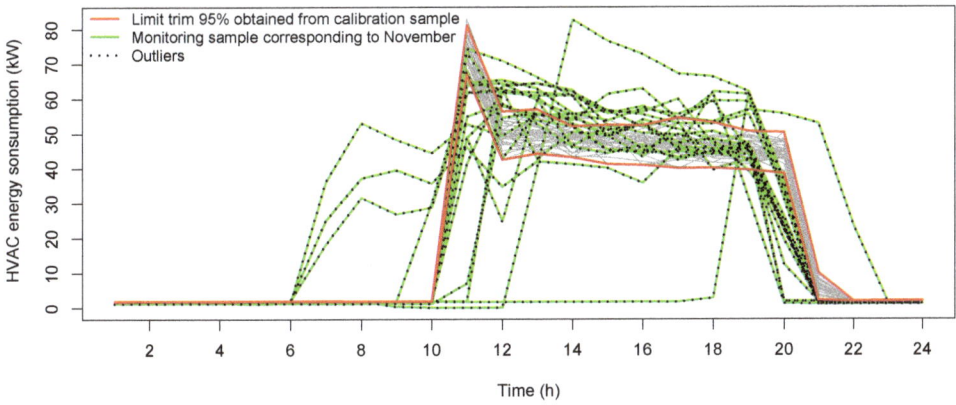

Figure 7. Reference (August and September) and monitoring samples corresponding to November.

Given that the process has changed, a new reference sample should be obtained and studied. The stabilization and monitoring process should begin for the next year, starting December. Figure **??** shows the rank control chart for the calibration (August and September) and monitoring samples of November. In practice, monitoring can be efficiently performed using the proposed control charts because you do not need to estimate the LCL through resampling procedures. Additionally, the rank control chart can be adapted for monitoring multivariate functional data, that is, when there are different types of curves that define the quality of a specific system or process simultaneously. For example, a direct case of application considers their first derivative along with the original curves.

The application of rank control charts also allows us to monitor the process by studying more than one functional variable. Particularly, it not only allows us to study the energy consumption curves but also the curves of daily temperature, daily relative humidity, daily CO_2 concentration, among others; they completely characterize the energy efficiency of the facilities and their thermal comfort and ambient air quality. Precisely, this procedure would allow us to perform a functional multivariate monitoring.

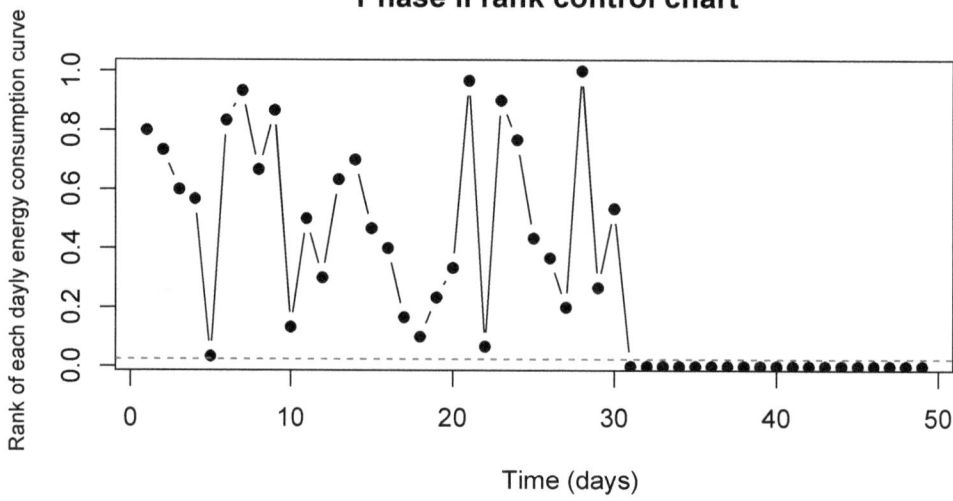

Figure 8. Rank control chart for Phase II of quality control. Both reference (corresponding to August and September) and monitoring (corresponding to November) samples are shown.

5. Simulation Study

The control chart performance is many times measured by observing its power $(1 - \beta)$. It is defined as the probability of identifying an out-of-control state when the process state is actually out-of-control [?]. Moreover, assuming that the process is actually under control, the type I error (α) or false alarm rate will be defined as the probability of detecting an out-of-control signal [?].

In the case of the process is under control, the probability of identifying an out-of-control observation should be small enough to prevent an unacceptable number of false alarms. Otherwise, if the process is effectively out of control, the power should be high enough to detect the process change as quickly as possible [?].

Another common index for measuring the performance of a control chart is Average Run Length (ARL), which is defined as the average number of observations plotted before a signal is out-of-control. The ARL is equal to $\frac{1}{p}$ (if we can assume that the signals are independent, that the run length distribution is geometric), where p is the probability of having an out-of-control signal [?].

To evaluate the performance of a control chart in Phase II, the ARL_0 and ARL_1 are often used, which are the average number of observations until the first out of control is detected, in cases where the process is really under control ($ARL_0 = \frac{1}{\alpha}$) or actually is not ($ARL_1 = \frac{1}{1-\beta}$) [?]. The ARL_1 should be at its lowest to increase the probability of quickly identifying events $(1 - \beta)$, power of a test that lead to the process being out-of-control [?].

Given that the F distribution is not known, a Monte Carlo simulation is designed to calculate the control charts power. The simulated scenarios allow us to estimate and compare the power of the control charts for different data depth measurements and for the case of independent and dependent functional data.

In this section, the performance of the control graphics proposed for Phases I and II will be evaluated. First, the simulation scheme designed in Febrero et al. [?] will be used to evaluate the performance of the control chart proposal for Phase I. Realizations of a Gaussian stochastic process have been proposed following the expression below [? ?]:

$$\mathcal{X}(t) = \mu(t) + \sigma(t) \cdot \epsilon(t), \qquad (3)$$

whereby $\sigma^2(t) = 0.5$ and

$$\mu(t) = \mathbf{E}(\mathcal{X}(t)) = 30t(1-t)^{3/2}, \qquad (4)$$

whereas $\epsilon(t)$ is a Gaussian process $\epsilon(t) \sim GP(\mathbf{0}, \Sigma)$ with $\mathbf{0}$ mean and variance-covariance matrix equal to

$$\mathbf{E}\left[\epsilon(t_i) \times \epsilon(t_j)\right] = e^{-\frac{|t_i - t_j|}{0.3}}.$$

Additionally, Reference [?] developed an alternative model to generate atypical curves, $\mu(t) = 30t^{3/2}(1-t)$. In Figure ??a, the two functional means are presented. The black curve accounts for the process mean without atypical curves, while the red curve is the mean of the process that generates the atypical curves.

The control charts proposed in this work have been designed to monitor the functional mean to detect two events—change in the mean of the process in terms of the magnitude and shape—which reveal that the process is not under control. For designing control charts for Phase II, it is assumed that the process is under control, that is, outliers are not detected. To generate simulation scenarios for each of these events, the following functional means have been considered:

- Mean of the model with a change in the magnitude:

$$\mu(t) = 30t(1-t)^{3/2} + \delta, \qquad (5)$$

by which δ denotes the change that goes from 0.4 to 2 in steps of 0.4.

- Mean of the model with a change in form:

$$\mu(t) = (1-\eta) \cdot 30t(1-t)^{3/2} + \eta \cdot 30t^{3/2}(1-t), \qquad (6)$$

where η is the change from 0.2 to 1 in steps of 0.2.

In Figure ??b, the green curve accounts for the functional mean of a process when there is a change in the magnitude ($\delta = 0.7$), while the curve of blue denotes the mean of a process when there is a change in the shape ($\eta = 0.3$).

In Febrero et al. [?], the functional data $\mathcal{X}_1, \ldots, \mathcal{X}_n$ denote realizations of a stochastic process $X(\cdot)$, assuming continuous trajectories in the $[a, b] = [0, 1]$ period and independence between the curves. However, simulation scenarios in which the simulated curves are defined by a variable degree of dependence have also been considered. This is because several practical applications of this type of chart are related to continuously monitored data with respect to time, forming functional time series, such as the curves of daily energy consumption in commercial areas. In this way, dependent curves are generated from the model $\tilde{Y}_i(t) = \mu(t) + \sigma(t) \cdot \tilde{\epsilon}(t)$, with $\tilde{\epsilon}(t) = \rho \cdot \tilde{\epsilon}_{i-1}(t) + (1-\rho) \cdot \epsilon_i(t)$, where ρ is the correlation measure between curves and $\sigma(t) = 0.5$ and both $\epsilon(t)$ and $\tilde{\epsilon}(t)$ are Gaussian processes [?].

In order to compare the results of the simulations in the scenarios defined by independence and dependence between the curves, the variance of ϵ is rescaled (we define the variance of the error $\tilde{\epsilon}$ to be one). Specifically, considering $\sigma_\epsilon^2 = \frac{(1-\rho^2)}{(1-\rho)^2} = \frac{(1+\rho)}{(1-\rho)}$, you have $\sigma_{\tilde{\epsilon}}^2 = 1$.

In Figure ??, different scenarios are presented considering the changes in the functional mean of the process, in the shape and magnitude, in the cases of independence and dependence between curves. The gray curves show the realizations of the process when it is under control (whose mean is the Equation (??)). However, the red curve in each graph accounts for the scenarios in which the presence of events that destabilize the process is considered, that is, the process is not under control. In Figure ??a,b, the cases of independence between curves and the presence of events defined by changes in the functional mean in terms of the magnitude and in shape, respectively, are shown.

However, Figure ??c,d show two cases defined by the presence of dependence between curves, including changes in the magnitude of the mean (panel c), with respect to its shape (panel d).

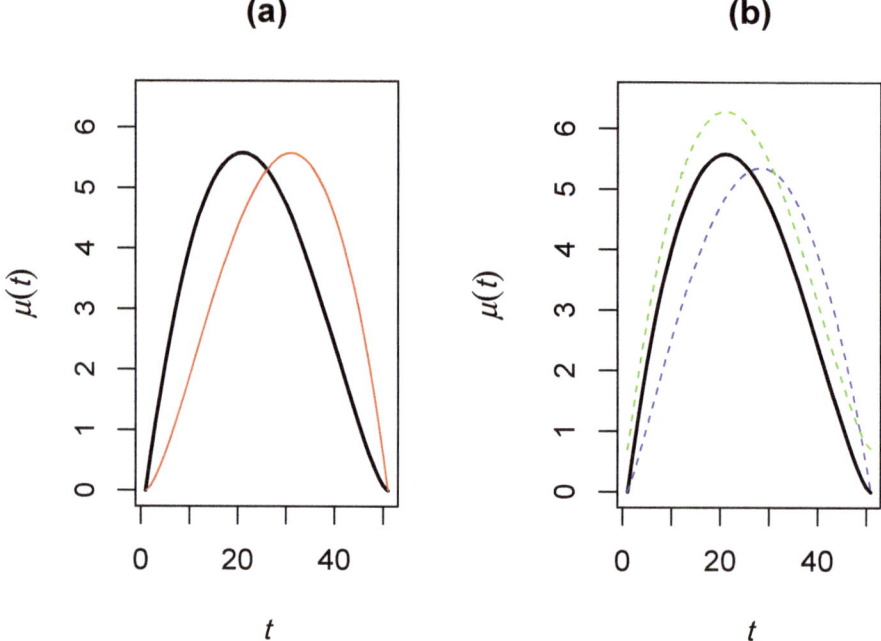

Figure 9. (a) Functional means and (b) changes in the shape and the magnitude in the mean of the process.

In the building energy efficiency domain, the out of control signals of energy consumption, temperature, CO_2 proportion and humidity, among others, can be defined by a change of shape and/or magnitude with respect to the under control signals analogous to those analyzed in the simulation study. The change in magnitude is related with a change in the scale of the studied process, for example, the increasing of energy consumption, temperature, humidity or concentration of CO_2 in all the hours corresponding of a specific day with respect to the otherwise normal pattern. This type of change is accounted by the addition of the δ term to the Equation (??) in order to obtain the Equation (??). Thus, the amount of that change is controlled by δ parameter. The green curve of Figure ?? accounts for an example of magnitude change. If we compare with the black one, we could realize that the two curves have the same shape and the only difference is the scale. On the other hand, changes in the shape of the curves are introduced by modifying the Equation (??) by the η parameter resulting in the Equation (??). In the energy efficiency domain, this type of change can be related to a change in the HVAC facilities programing (changes in temperature regulation, changes in the time schedule), in a failure of the HVAC in just one interval of the day, an extremely high or low level of occupation in a building and extreme changes in the weather, among other causes. The proposed simulation study has performed taking into account the specific domain where the FDA control chart approach is applied, namely energy efficiency in buildings. Thus, the shape of these types of profiles is similar to that corresponding to CO_2, temperature, energy consumption and humidity curves in buildings. In order to measure the performance of the proposed control charts for very different types of profiles, new studies may be necessary.

In the following section, a simulation study is performed to determine conditions under which it can be verified that the smooth bootstrap procedure works when there is independence and dependence between curves.

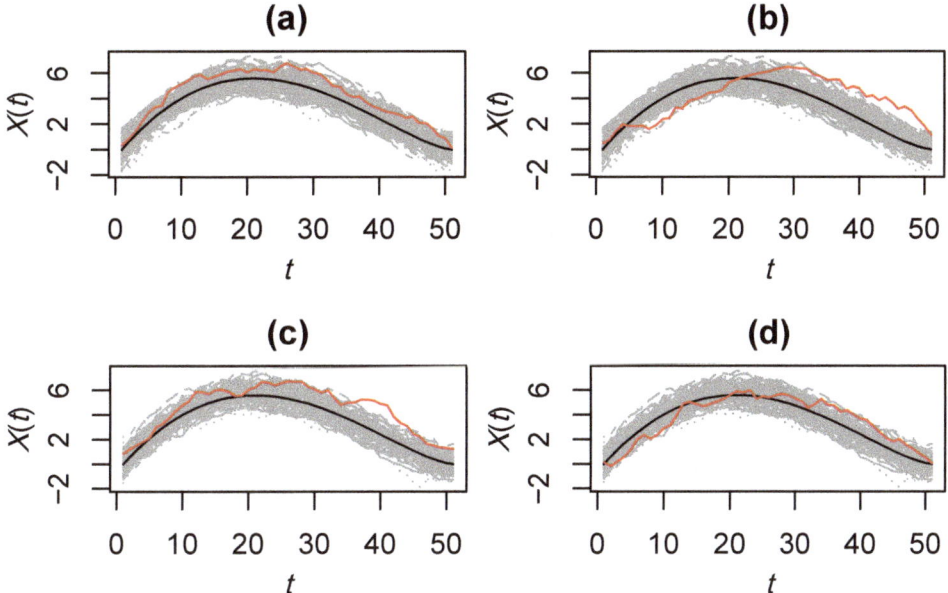

Figure 10. Scenarios in which independence between curves is studied. Changes in the functional mean with respect to its magnitude (**a**) and shape (**b**) are shown. In the case of dependence, panels (**c**) and (**d**) show the simulation scenarios in which changes in the magnitude and shape, respectively, are observed in the functional mean.

5.1. Measurement and Comparison of the Performance of the Control Chart Proposed for Phase I

The performance of the control chart is estimated and compared from the generation of calibrated samples of size 50 and 100 (curves). For each sample, different functional depth measurements described in Section ?? are calculated and the outlier detection robust procedures (weighted and trimmed) are applied for the estimation of type I error when the process is under control and the power of the test when the process is out-of-control. For the estimation of type I error, each scenario is replicated 1000 times ($n = 50, 100$, assuming independence and dependence between curves). When the power of the test is estimated, in each scenario (assuming independence and dependence), a curve within the alternative hypothesis is generated; this procedure is also repeated 1000 times.

Following the scheme described in Reference [?], curves observed at equidistant points are considered; the number of points that define each curve is 51 in the interval $[0,1]$. From 1000 resamples ($B = 1000$) and with a 2.5% trimming procedure (removing less deep curves), a smoothing bootstrap procedure defined by a smoothing factor $\gamma = 0.05$ is applied to estimate the $C = 0.01$ quantile representing the LCL.

First, a simulation study is performed to estimate and compare the type I error ($\alpha = 0.01$ is fixed) of the proposed control chart, assuming scenarios with independence and dependence between curves. Subsequently, a similar study is carried out to estimate and evaluate the power of the control chart to detect out-of-control signals in different situations (independence, dependence, different sample size and a change in the shape or magnitude).

In Table ??, the results of the estimation of the false alarm rate (type I error) in the independence scenario are shown. The average of the percentage of false out-of-control signals (type I error) detected by the procedure shown above are very close to the nominal 1% for the two considered sample sizes. Furthermore, it can be observed that, when n increases, the type I error percentages are closer to the nominal level. In general, for a sample of size $n = 100$, the results of applying the weighted method are closer to α, especially when using the mode depth measurement. The results obtained in the simulations are similar to those presented in Reference [?].

Table 1. Estimation of the false alarm rate (%) for the case of independence between curves using a nominal type I error of 1%.

n	Method	FM	RP	Mode
50	Weighted	1.94	1.89	1.49
	Trimmed	1.34	1.95	1.36
100	Weighted	1.55	1.75	1.25
	Trimmed	1.67	2.33	1.76

In any process, type I error increases the production cost. Hence, it is essential not to overestimate this error rate when managing the quality.

Table ?? shows the results of the simulation to evaluate the ability of the control chart to detect a change in the shape or magnitude of the functional mean of the process through the estimation of its power $(1 - \beta)$. The percentage of out-of-control signals (outliers) correctly detect when the population defined by the Equation (??) is contaminated with curves belonging to the M_1 model (Equation (??)) and M_2 model (Equation (??)); it is denoted by p_c; however, the percentage of false alarms (false states out of control) is p_f. These parameters have been estimated, in all the scenarios assuming the independence of curves, using the average of the corresponding empirical values, $\hat{p}_c$ and $\hat{p}_f$.

Table ?? shows that a better performance is achieved when the curves of model M_1 (where changes in the magnitude of the proposed control chart are simulated) are studied. Precisely, $\hat{p}_f$ and $\hat{p}_c$ are closer to the nominal α and $(1 - \beta)$. When identifying changes in the shape of the process average, M_2, the mode depth provides the highest percentages of correctly detected out-of-control signals. However, in the case of the M_1 model, the use of RP depth provides percentages of the detection of the true out-of-control states lower than those corresponding to the use of FM and mode depths. With respect to a robust method for outlier detection, the performance is similar in all the scenarios. However, an exception is the case wherein the RP depth is used; it reveals the low performance of the control chart in detecting observations corresponding to actual out-of-control states.

Briefly, the detection rate of false out-of-control signals for the independence scenario is close to 1%. However, when using the trimmed method, the detection rate of false out-of-control signals is overestimated but this percentage decreases when the sample size increases.

The results of the false alarm rate (type I error) for scenarios defined by dependence between curves are shown in Table ??. It is important to note that, for different values of ρ, very similar results with respect to the independence scenarios have been obtained. Precisely, the average of the percentages of false out-of-control signals are close to the nominal 1% in the two studied sample sizes. Additionally, when n increases, the type I error percentages are closer to the nominal level. However, some differences are observed when the RP data depth measure is used to develop the control chart. In this case, there is an overestimation of the percentage of false out-of-control signals.

Table 2. Percentages of $\hat{p}_c$ and $\hat{p}_f$ for the cases of curves simulated with M_1 (Equation (??)) and M_2 (Equation (??)) models, assuming independence between curves.

Stage	Displays	Depth	Method	δ 0.4 $\hat{p}_f$	$\hat{p}_c$	0.8 $\hat{p}_f$	$\hat{p}_c$	1.2 $\hat{p}_f$	$\hat{p}_c$	1.6 $\hat{p}_f$	$\hat{p}_c$	2 $\hat{p}_f$	$\hat{p}_c$
M_1	50	FM	Weighted	1.80	8.30	1.63	27.80	1.47	56.50	1.48	85.30	1.60	95.70
			Trimmed	1.25	7.70	1.19	26.80	1.19	56.00	1.22	83.00	1.20	95.50
		RP	Weighted	1.88	6.80	1.68	20.10	1.49	45.10	1.42	71.90	1.42	87.70
			Trimmed	1.88	7.00	1.89	23.20	2.08	49.30	2.14	75.70	2.20	91.90
		Mode	Weighted	1.41	5.70	1.27	19.90	1.09	45.10	1.08	76.70	1.17	94.30
			Trimmed	1.33	6.80	1.32	22.20	1.33	48.60	1.38	77.50	1.39	95.20
	100	FM	Weighted	1.51	5.70	1.40	23.30	1.34	53.50	1.34	82.00	1.38	96.20
			Trimmed	1.63	7.40	1.51	25.90	1.32	56.10	1.13	83.60	1.00	96.70
		RP	Weighted	1.68	5.00	1.57	18.60	1.50	45.40	1.45	71.40	1.52	90.10
			Trimmed	2.28	7.40	2.22	23.70	2.23	54.20	2.13	78.70	2.08	93.50
		Mode	Weighted	1.22	4.20	1.15	17.90	1.08	46.80	1.06	78.90	1.12	94.70
			Trimmed	1.70	5.80	1.62	22.90	1.46	54.20	1.28	83.50	1.13	95.90
				η 0.2		0.4		0.6		0.8		1	
M_2	50	FM	Weighed	1.91	1.50	1.85	3.40	1.70	11.00	1.46	26.30	1.25	55.00
			Trimmed	1.28	1.90	1.23	4.10	1.19	11.60	1.13	26.70	1.18	50.60
		RP	Weighted	1.94	2.40	1.90	3.30	1.75	8.00	1.64	16.90	1.42	32.20
			Trimmed	1.89	2.30	1.83	4.60	1.73	10.10	1.84	22.50	1.92	41.40
		Mode	Weighted	1.43	2.80	1.35	10.70	1.19	31.00	1.06	65.30	1.12	91.70
			Trimmed	1.33	4.00	1.31	12.30	1.31	33.70	1.42	68.00	1.48	92.10
	100	FM	Weighted	1.53	2.00	1.51	2.70	1.45	12.20	1.35	27.90	1.25	53.00
			Trimmed	1.66	2.50	1.63	4.30	1.57	14.40	1.51	33.40	1.43	58.00
		RP	Weighted	1.69	2.70	1.66	4.10	1.63	8.20	1.55	17.30	1.48	34.70
			Trimmed	2.25	2.90	2.19	5.60	2.15	13.60	2.15	25.40	2.21	45.50
		Mode	Weighted	1.22	3.30	1.18	11.90	1.12	34.10	1.07	66.80	1.09	89.60
			Trimmed	1.72	4.60	1.65	15.70	1.57	41.70	1.41	72.70	1.22	92.90

Table 3. Results for the scenarios with dependence between curves. The estimates of false alarm rate (type I error) for values of ρ between 0.3 and 0.7 are shown.

ρ	n	Method	FM	RP	Mode
0.3	50	Weighted	1.85	1.88	1.41
		Trimmed	1.31	1.90	1.33
	100	Weighted	1.54	1.71	1.20
		Trimmed	1.65	2.23	1.71
0.5	50	Weighted	1.80	1.81	1.28
		Trimmed	1.26	1.82	1.29
	100	Weighted	1.46	1.68	1.15
		Trimmed	1.61	2.21	1.65
0.7	50	Weighted	1.66	1.73	0.93
		Trimmed	1.25	1.74	1.06
	100	Weighted	1.42	1.61	0.99
		Trimmed	1.56	2.15	1.57

Tables ??–?? show the results of the empirical estimation of p_c and p_f, assuming different values of ρ (from 0.3 to 0.7). The power (estimated by $\hat{p}_c$) of the control chart proposed for the model M_1 (Equation (??)) performs better when the weighted method is applied and if the sample size is increased. It is also observed that the performance of the control chart tends to be the same, independent of the type of data depth measurement used. Certainly, the performance of control charts in detecting real changes in the process, related to differences in the shape and mean, is better when the mode depth is used.

Table 4. Empirical values of $\hat{p}_f$ and $\hat{p}_c$, with $\rho = 0.3$ (assuming dependence between curves).

Model	Sample Size	Depth	Method	δ	0.4		0.8		1.2		1.6		2	
					$\hat{p}_f$	$\hat{p}_c$	$\hat{p}_f$	$\hat{p}_c$	$\hat{p}_f$	$\hat{p}_c$	$\hat{p}_f$	$\hat{p}_c$	$\hat{p}_f$	$\hat{p}_c$
M_1	50	FM	Weighted		1.58	8.90	1.22	33.60	0.98	69.80	1.07	91.70	1.19	99.30
			Trimming		1.12	9.20	0.75	26.60	0.37	47.40	0.09	54.30	0.02	49.20
		RP	Weighted		1.75	7.10	1.38	26.30	0.97	53.80	0.74	77.60	0.67	90.70
			Trimmed		1.72	7.60	1.42	25.90	0.89	45.90	0.49	60.90	0.20	68.70
		Mode	Weighted		1.21	5.80	0.90	24.90	0.67	59.80	0.70	88.30	0.86	98.50
			Trimmed		1.21	6.40	0.89	24.00	0.50	49.40	0.19	73.50	0.04	89.80
	100	FM	Weighted		1.47	7.30	1.23	29.60	1.07	67.80	1.13	92.30	1.24	99.20
			Trimmed		1.53	7.60	1.29	31.20	1.04	68.70	0.96	91.30	0.92	99.10
		RP	Weighted		1.69	5.50	14.67	24.40	1.21	58.00	1.06	83.90	1.02	95.90
			Trimmed		2.19	8.80	1.99	27.40	1.88	64.80	1.92	89.30	1.85	97.30
		Mode	Weighted		1.17	4.60	0.98	23.00	0.86	62.40	0.89	91.00	1.01	98.50
			Trimming		1.62	6.90	1.38	27.90	1.15	67.10	1.07	91.70	1.06	98.90
				η	0.2		0.4		0.6		0.8		1	
M_2	50	FM	Weighted		1.75	1.30	1.61	1.95	1.31	7.70	0.94	16.85	0.64	26.05
			Trimmed		1.25	1.40	1.15	2.35	0.97	7.70	0.73	15.95	0.51	20.50
		RP	Weighted		1.86	1.20	1.75	3.00	1.43	5.50	1.24	10.85	0.96	16.55
			Trimmed		1.79	1.65	1.68	3.40	1.58	7.15	1.46	13.10	1.16	19.40
		Mode	Weighted		1.31	3.70	1.08	16.00	0.78	44.90	0.62	79.30	0.81	96.90
			Trimmed		1.22	4.70	1.11	17.50	0.74	42.20	0.29	65.00	0.07	82.50
	100	FM	Weighted		1.53	0.90	1.46	2.05	1.30	7.45	1.07	19.50	0.92	32.35
			Trimmed		1.65	1.20	1.57	2.70	1.44	9.40	1.22	22.75	1.12	35.35
		RP	Weighted		1.76	0.60	1.73	2.20	1.57	5.10	1.41	12.10	1.27	21.80
			Trimmed		2.23	1.10	2.18	3.30	2.09	8.25	1.91	17.45	1.97	29.35
		Mode	Weighted		1.17	3.55	1.07	14.70	0.91	47.20	0.86	83.10	0.98	97.50
			Trimmed		1.64	4.70	1.48	20.50	1.25	52.60	1.13	85.10	1.12	98.20

Table 5. Empirical values of $\hat{p}_f$ and $\hat{p}_c$, with $\rho = 0.5$ (assuming dependence between curves).

Model	Sample Size	Depth	Method	δ	0.4		0.8		1.2		1.6		2	
					$\hat{p}_f$	$\hat{p}_c$	$\hat{p}_f$	$\hat{p}_c$	$\hat{p}_f$	$\hat{p}_c$	$\hat{p}_f$	$\hat{p}_c$	$\hat{p}_f$	$\hat{p}_c$
M_1	50	FM	Weighted		1.46	11.50	1.04	43.90	0.84	80.80	1.06	97.40	1.14	99.90
			Trimmed		1.02	10.80	0.63	33.40	0.23	50.70	0.28	53.80	0.00	45.60
		RP	Weighted		1.68	9.10	1.21	31.10	0.80	63.00	0.65	85.60	0.60	94.60
			Trimmed		1.73	9.60	1.26	30.70	0.66	52.60	0.28	66.40	0.11	68.80
		Mode	Weighted		1.10	7.50	0.76	34.40	0.55	73.00	0.69	95.20	0.82	99.90
			Trimmed		1.05	8.10	0.72	31.40	0.32	62.60	0.06	86.50	0.01	96.40
	100	FM	Weighted		1.20	10.00	1.08	39.60	1.01	79.40	1.10	97.30	1.18	99.80
			Trimmed		1.30	10.00	1.18	41.60	0.96	80.20	0.89	97.00	0.84	99.80
		RP	Weighted		1.40	10.00	1.39	32.30	1.09	69.20	0.97	90.60	0.97	97.90
			Trimmed		1.80	10.00	1.93	38.20	1.84	77.10	1.80	94.70	1.73	98.80
		Mode	Weighted		0.90	10.00	0.86	32.10	0.79	75.80	0.91	96.00	0.98	99.90
			Trimmed		1.30	10.00	1.30	38.30	1.10	79.40	1.06	97.20	1.02	99.90
				η	0.2		0.4		0.6		0.8		1	
M_2	50	FM	Weighted		1.63	1.40	1.43	3.55	1.02	10.80	0.68	21.60	0.50	33.10
			Trimmed		1.16	1.50	1.02	3.60	0.80	11.20	0.54	18.00	0.32	21.70
		RP	Weighted		1.89	1.75	1.65	3.30	1.32	7.30	0.98	14.30	0.75	22.55
			Trimmed		1.79	2.30	1.62	3.80	1.54	9.05	1.26	16.40	0.95	22.90
		Mode	Weighted		1.13	5.40	0.86	22.60	0.56	60.50	0.59	91.50	0.85	99.80
			Trimmed		1.10	6.30	0.83	23.40	0.46	54.00	0.11	77.20	0.02	93.80
	100	FM	Weighted		0.01	0.01	0.01	0.03	1.13	11.90	0.90	28.25	0.77	38.90
			Trimmed		0.02	0.01	0.01	0.04	1.28	15.25	1.11	30.90	1.04	39.90
		RP	Weighted		0.02	0.01	0.02	0.03	1.42	7.65	1.26	16.65	1.08	27.55
			Trimmed		0.03	0.02	0.02	0.05	2.02	12.60	1.91	23.25	1.96	35.00
		Mode	Weighted		0.01	0.05	0.01	0.23	0.82	63.00	0.87	92.50	0.99	99.80
			Trimmed		0.02	0.07	0.01	0.28	1.21	68.40	1.12	93.40	1.12	99.60

Table 6. Empirical values of $\hat{p}_f$ and $\hat{p}_c$, with $\rho = 0.7$ (assuming dependence between curves).

			δ	0.4		0.8		1.2		1.6		2	
Model	Sample Size	Depth	Method	$\hat{p}_f$	$\hat{p}_c$	$\hat{p}_f$	$\hat{p}_c$	$\hat{p}_f$	$\hat{p}_c$	$\hat{p}_f$	$\hat{p}_c$	$\hat{p}_f$	$\hat{p}_c$
M_1	50	FM	Weighted	1.32	16.00	0.87	57.70	0.85	91.40	1.01	99.30	1.01	100.00
			Trimmed	0.87	13.40	0.42	40.30	0.07	53.60	0.00	47.50	0.00	35.80
		RP	Weighted	1.41	13.00	0.97	43.40	0.61	76.70	0.56	92.00	0.53	98.00
			Trimmed	1.51	14.00	0.97	40.70	0.39	61.40	0.14	68.20	0.04	71.30
		Mode	Weighted	0.71	14.00	0.44	52.20	0.41	88.90	0.51	99.30	0.53	100.00
			Trimmed	0.82	14.30	0.43	47.40	0.09	80.20	0.01	96.00	0.00	99.90
	100	FM	Weighted	1.29	12.00	0.99	54.90	1.05	90.40	1.14	99.50	1.13	100.00
			Trimmed	1.36	13.80	1.08	56.00	0.89	90.20	0.84	99.20	0.76	100.00
		RP	Weighted	1.50	9.90	1.20	48.30	0.97	80.80	0.93	96.40	0.96	99.50
			Trimmed	2.00	13.90	1.80	53.10	1.80	87.30	1.72	98.30	1.60	99.90
		Mode	Weighted	0.90	9.80	0.68	51.00	0.69	90.00	0.81	99.40	0.83	99.90
			Trimmed	1.43	13.40	1.14	58.10	1.01	92.40	0.97	99.60	0.92	99.90
			η	0.2		0.4		0.6		0.8		1	
M_1	50	FM	Weighted	1.52	1.80	1.21	6.05	0.76	17.90	0.49	30.70	0.41	39.95
			Trimming	1.07	2.10	0.85	6.15	0.56	15.30	0.31	22.30	0.18	24.05
		RP	Weighted	1.62	2.10	1.36	5.25	0.98	11.40	0.68	19.95	0.56	28.30
			Trimmed	1.59	2.40	1.41	5.80	1.19	13.85	0.92	21.05	0.70	28.15
		Mode	Weighted	0.75	8.60	0.47	39.90	0.37	82.80	0.50	99.10	0.61	100.00
			Trimmed	0.87	10.30	0.51	38.30	0.14	73.10	0.01	93.80	0.00	99.30
	100	FM	Weighted	2.02	0.00	1.24	6.35	0.99	21.20	0.81	36.45	0.75	45.10
			Trimmed	3.03	0.00	1.36	7.80	1.16	24.30	1.05	37.95	1.03	44.55
		RP	Weighted	3.69	0.00	1.50	5.25	1.25	13.00	1.08	24.50	0.99	34.85
			Trimmed	6.28	0.00	1.98	7.20	1.84	18.50	1.89	31.80	1.89	41.60
		Mode	Weighted	1.63	0.00	0.72	37.90	0.69	82.70	0.80	98.60	0.90	100.00
			Trimmed	3.48	0.00	1.24	46.70	1.10	87.10	1.05	99.00	1.04	100.00

With respect to the false out-of-control rate $\hat{p}_f$, in the scenarios corresponding to the use of M_1 model, when the trimmed method is also used, a lower rate is obtained. In the case of the M_2 model, there are similar results on the scenarios defined by independence between curves, that is, the $\hat{p}_f$ is lower when the trimmed method for outlier detection is used.

In Reference [?], new methods for the detection of outliers were proposed for the case in which there is dependence between curves. From the simulation studies carried out in this study, at different degrees of dependence, we can say that the outlier detection method proposed in Reference [?] was relatively robust against the presence of dependence between curves. The simulation study performed in this section supports the results obtained in the work in Reference [?] and, in conclusion, justifies the use of this method within the new control charts proposed for Phase I, even in scenarios with dependence between curves.

Although the application of the weighed outlier detection method to Mode data depth has generally provided Phase I control charts with best performance, if the false alarm rate of Tables ?? and ?? are observed, the use of trimmed outlier detection method tends to provide values of $\hat{p}_f$ slightly closer to $\alpha = 1\%$ (with respect to the weighted method) when the process is actually under control, the curves are independent and the sample is relatively small ($n = 50$). In addition, if the process is out of control, the curves are independent and the outliers are generated by the Equation (??) (changes in magnitude), the trimmed method applied to FM data depth provide a $\hat{p}_f$ close to $\alpha = 1\%$ and the highest $\hat{p}_c$, as shown in Table ??. In all the remaining scenarios, the use of weighed method applied to Mode data depth tends to provide the closest to $\alpha\%$ $\hat{p}_f$ and the highest $\hat{p}_c$ (see Tables ??–??).

5.2. Measurement and Comparison of the Performance of the Control Chart Proposed for Phase II

For Phase II, the monitoring stage, the use of the rank control chart has been proposed. The application of the rank control chart allows simultaneous monitoring of changes in the mean and variability of a process. In the functional case, in order to calculate the rank statistic, the functional FM, RP and mode depths are used.

An $ARL_0 = \frac{1}{\alpha=0.025}$ (the monitoring sample is assumed under control) is assumed to evaluate the performance of the control chart. The power of the control chart is estimated and compared for an under-control process, based on the generation of a calibration sample of size $n = 50$ by a Monte Carlo procedure.

Following the simulation scheme of Phase I, curves observed at equidistant points are assumed; they are composed of 51 points at $[0, 1]$ interval. A smoothed bootstrap with a smoothing factor $\gamma = 0.05$, 1000 resamples ($B = 1000$), using a 2.5% trimmed procedure (removing the shallowest curves), is applied to estimate and compare the power of the control chart to detect out-of-control signals when a significance level of $\alpha = 0.025$ is assumed. Additionally, in the same way as in Phase I, the simulation of scenarios with independence and dependence between curves are assumed.

Table ?? shows the estimates of the power (%) of the control charts for the scenario of independence between curves, whereas Table ?? shows power of the control chart for the scenario with dependence. In both cases, the ability of the control chart to detect a change in the magnitude of the functional mean of the process is evaluated by the estimation of its power $(1 - \beta)$.

Table 7. Power of the control chart, $1 - \beta$, for the case M_1 (Equation (??)) model in the scenario of independence between curves.

δ	FM	RP	Mode
0.5	14.5	14.8	14.8
1	47.6	47.2	48.0
1.5	83.3	80.8	83.0
2	97.8	97.5	97.7

From the results of the Table ??, any depth measure can be used to detect a shift in the process mean, since the same performance, in terms of power, is obtained.

Table 8. Power of the control chart, $1 - \beta$, for the scenarios defined by the M_1 (??) model, assuming dependence between curves.

ρ	δ	FM	RP	Mode
0.3	0.5	20.28	23.73	21.84
	1	62.95	64.75	64.06
	1.5	92.98	92.33	93.64
	2	99.56	99.16	99.75
0.5	0.5	25.59	29.41	28.41
	1	73.84	75.12	76.24
	1.5	97.15	96.56	97.94
	2	99.91	99.77	99.98
0.7	0.5	35.47	40.23	43.55
	1	86.14	86.87	90.38
	1.5	99.48	99.11	99.78
	2	100.00	99.99	100.00

The results of the detection of a shift in the process mean are shown in Table ??. A similar performance is observed when using any depth measure for different values of ρ. Apparently, the control chart for Phase II is robust against the existence of dependence between curves.

As observed in the simulation study and in the analysis of the case study with real data, the present proposal of control charts for functional data, including Phase I and II control charts, can be useful to detect anomalies in diverse scenarios. In the case of its application to real data, the set of proposed techniques is being examined for implementation in the web platform Σqus and for its use by the company Nerxus for detecting false alarms in facilities in commercial areas. The present control chart methodology can be used for control tasks, monitoring, anomaly detection and continuous improvement in diverse industrial processes, monitoring of environmental variables, chemical industry and, in general, any process involving continuous monitoring of functional data over time.

Regarding the use of our methodology in more complex case studies defined by different operation modes, the application of the multi-modelling framework methodology in combination with our proposal could be useful. Indeed, in the building energy efficiency domain, there are many different operation modes of installations, each one defined by a specific operation pattern. Namely, the HVAC installations can be operated in heating or ventilation modes (there are even different modes within ventilation or heating). The automatic classification of each profile in the corresponding profile pattern could be very useful in the building energy efficiency field and previously to the application of our control chart proposal for Phase I and Phase II. With respect to the work of Grasso et al. [?], it is also interesting to mention that the proposed profile monitoring control charting scheme is that based on functional PCA and described in Colosimo and Pacella [?].

6. Conclusions

A new alternative of control charts has been proposed when CTQ variables of the process are functional. The proposal includes alternatives to develop the the Phase I and II control charts for stabilizing and monitoring the processes, respectively. In order to develop Phase I control charts based on functional data, outlier detection methods are used based on a method of smooth bootstrap resampling and the depth calculation of functional data is proposed. However, in order to implement Phase II, the use of rank-type nonparametric control charts based on the concept of functional data depth is proposed. This Phase II control chart is directly estimated assuming that the asymptotic distribution of the rank statistic is a uniform distribution. The application of the control charts to the two-process control phases and the development of a new graphic tool for visualizing functional data (including an envelope with 95% of the deepest curves that facilitate the identification of the assignable cause of each anomaly) give rise to the proposed methodology. It has been successfully applied in real case studies belonging to the framework of anomaly detection in building energy efficiency. Additionally, a simulation study is conducted to measure the performance (as the percentage of rejection when the null hypothesis is not met) of the control charts, depending on the functional data depth used, the sample size, the presence of dependence between curves and the use of different FDA procedures for outlier detection.

In the simulation study, the use of different types of functional depths has been compared to develop Phase II of the proposed control chart. In case of the univariate functional data (single type of curves), for the three scenarios, a better performance is obtained with the mode depth measurement combined with the weighted outlier detection method and moderately large samples. Additionally, one of the final observations of the simulation study is that the control chart methodology is robust against the presence of dependence between curves. Thus, this alternative tool can be applied to the framework of continuously monitored data streams.

Generally, the authors recommend using the weighed method and the Mode functional data depth for the case of Phase I taking into account the values of $\hat{p}_f$ and $\hat{p}_c$. Thus, when the Phase I control chart is evaluated, both weighted method and Mode data depth are generally the best options in those scenarios defined by under control assumption and even in those where out of control curves are simulated. In the latter, when the change in magnitude or shape is very small, the corresponding $\hat{p}_c$ tends to be not higher to those obtained by the use of other combinations of data depth measure and outlier detection method. Nevertheless, when the change in magnitude (δ) or shape (η) increases, the power of the combination of weighted method and Mode depth tends to be higher than those corresponding to the other combinations. Moreover, regarding the Phase II control chart and taking into account the higher values of power estimates included in Tables ?? and ??, we also recommend the use of Mode data depth for Phase II control charts.

The present proposal has been verified by its application in a real case study dealing with the detection of energy efficiency anomalies in buildings. Specifically, all the previously identified real anomalies (by the maintenance personnel) have also been successfully identified by the application of this functional approximation of control charts for Phases I and II of process control. Additionally, the proposed graphical tool helps to intuitively identify the assignable causes corresponding to each anomaly.

This procedure can be used in different industrial and scientific domains in which the control procedures are defined by functional CTQ variables.

Author Contributions: Conceptualization, S.N., M.F. and J.T.-S.; methodology, M.F., R.F.-C. and S.N.; software, M.F. and R.F.-C.; validation, M.F., R.F.-C. and J.T.-S.; formal analysis, M.F., R.F.-C., S.N. and J.T.-S.; investigation, M.F., R.F.-C., S.N., J.T.-S., S.Z. and P.R.; resources, S.N., S.Z. and P.R.; data curation, S.Z. and P.R.; writing—original draft preparation, M.F., J.T.-S., S.N. and R.F.-C.; writing—review and editing, J.T.-S., R.F.-C., M.F., S.N. and S.Z.; visualization, Miguel Flores, R.F.-C. and S.N.; supervision, S.N., J.T.-S. and R.F.-C.; project administration, S.N.; funding acquisition, S.Z. and S.N. All authors have read and agreed to the published version of the manuscript.

Funding: This study has been funded by the eCOAR project (PC18/03) of CITIC. The work of Salvador Naya, Javier Tarrío-Saavedra, Miguel Flores and Rubén Fernández-Casal has been supported by MINECO grants MTM2014-52876-R, MTM2017-82724-R, the Xunta de Galicia (Grupos de Referencia Competitiva ED431C-2016-015, and Centro Singular de Investigación de Galicia ED431G/01 2016-19), through the ERDF. The research of Miguel Flores has been partially supported by Grant PII-DM-002-2016 of Escuela Politécnica Nacional of Ecuador.

Acknowledgments: The authors strongly thank, on the one hand, Fridama, Σqus, and Nerxus companies, and on the other hand CITIC, Campus Industrial and MODES group their valuable help and support.

Conflicts of Interest: The authors declare no conflict of interest. The founding sponsors had no role in the design of the study; in the collection, analyses, or interpretation of data; in the writing of the manuscript, or in the decision to publish the results.

References

- Lu, C.W.; Reynolds, M.R, Jr. EWMA control charts for monitoring the mean of autocorrelated processes. *J. Qual. Technol.* **1999**, *31*, 166. [CrossRef]
- Alwan, L.C.; Roberts, H.V. Time-series modeling for statistical process control. *J. Bus. Econ. Stat.* **1988**, *6*, 87–95.
- Qiu, P.; Zou, C.; Wang, Z. Nonparametric profile monitoring by mixed effects modeling. *Technometrics* **2010**, *52*, 265–277. [CrossRef]
- Shiau, J.J.; Huang, H.L.; Lin, S.H.; Tsai, M.Y. Monitoring nonlinear profiles with random effects by nonparametric regression. *Commun. Stat. Methods* **2009**, *38*, 1664–1679. [CrossRef]
- Noorossana, R.; Saghaei, A.; Amiri, A. *Statistical Analysis of Profile Monitoring*; John Wiley & Sons: Hoboken, NJ, USA, 2011; Volume 865.
- Chipman, H.; MacKay, R.; Steiner, S. Comment on Nonparametric profile monitoring by mixed effects modeling. *Technometrics* **2010**, *52*, 280–283. [CrossRef]
- Lavin, A.; Ahmad, S. Evaluating Real-Time Anomaly Detection Algorithms–The Numenta Anomaly Benchmark. In Proceedings of the 2015 IEEE 14th International Conference on Machine Learning and Applications (ICMLA), Miami, FL, USA, 9–11 December 2015; pp. 38–44.
- Kroll, B.; Schaffranek, D.; Schriegel, S.; Niggemann, O. System modeling based on machine learning for anomaly detection and predictive maintenance in industrial plants. In Proceedings of the 2014 IEEE Emerging Technology and Factory Automation (ETFA), Barcelona, Spain, 16–19 September 2014; pp. 1–7.
- Ahmad, S.; Lavin, A.; Purdy, S.; Agha, Z. Unsupervised real-time anomaly detection for streaming data. *Neurocomputing* **2017**, *262*, 134–147. [CrossRef]
- Meshram, A.; Haas, C. Anomaly detection in industrial networks using machine learning: A roadmap. In *Machine Learning for Cyber Physical Systems*; Springer: Berlin/Heidelberg, Germany, 2017; pp. 65–72.
- Basu, S.; Meckesheimer, M. Automatic outlier detection for time series: An application to sensor data. *Knowl. Inf. Syst.* **2007**, *11*, 137–154. [CrossRef]
- Mosallam, A.; Medjaher, K.; Zerhouni, N. Nonparametric time series modelling for industrial prognostics and health management. *Int. J. Adv. Manuf. Technol.* **2013**, *69*, 1685–1699. [CrossRef]
- Talagala, P.D.; Hyndman, R.J.; Smith-Miles, K.; Kandanaarachchi, S.; Muñoz, M.A. Anomaly detection in streaming nonstationary temporal data. *J. Comput. Graph. Stat.* **2019**, pp. 1–28. [CrossRef]

- Hyndman, R.J.; Liu, X.A.; Pinson, P. Visualizing big energy data: Solutions for this crucial component of data analysis. *IEEE Power Energy Mag.* **2018**, *16*, 18–25. [CrossRef]
- Golshan, M.; MacGregor, J.F.; Bruwer, M.J.; Mhaskar, P. Latent Variable Model Predictive Control (LV-MPC) for trajectory tracking in batch processes. *J. Process Control* **2010**, *20*, 538–550. [CrossRef]
- Kourti, T.; MacGregor, J.F. Multivariate SPC methods for process and product monitoring. *J. Qual. Technol.* **1996**, *28*, 409–428. [CrossRef]
- Ferrer-Riquelme, A. 1.04-Statistical Control of Measures and Processes. In *Comprehensive Chemometrics*; Elsevier: Amsterdam, The Netherlands, 2009; pp. 97–126.
- Colosimo, B.M.; Pacella, M. A comparison study of control charts for statistical monitoring of functional data. *Int. J. Prod. Res.* **2010**, *48*, 1575–1601. [CrossRef]
- Megahed, F.; Jones-Farmer, L. Statistical Perspectives on "Big Data". In *Frontiers in Statistical Quality Control 11*; Springer: Berlin, Germany, 2015; pp. 29–47.
- Woodall, W.; Montgomery, D. Some current directions in the theory and application of statistical process monitoring. *J. Qual. Technol.* **2014**, *46*, 78. [CrossRef]
- Sheu, S.H.; Ouyoung, C.W.; Hsu, T.S. Phase II statistical process control for functional data. *J. Stat. Comput. Simul.* **2013**, *83*, 2144–2159. [CrossRef]
- García, D.R. Cartas de Control Para Datos Funcionales. Master's Thesis, Centro de Investigación en Matemáticas (CIMAT), Guanajuato, Mexico, 2011.
- Rodrigo, O.P. Monitoreo de Datos Funcionales. Master's Thesis, Centro de Investigación en Matemáticas (CIMAT), Guanajuato, Mexico, 2013.
- Flores, M.; Tarrio-Saavedra, J.; Fernandez-Casal, R.; Naya, S. Functional extensions of Mandel's h and k statistics for outlier detection in interlaboratory studies. *Chemom. Intell. Lab. Syst.* **2018**, *176*, 134–148. [CrossRef]
- Flores, M.; Fernández-Casal, R.; Naya, S.; Tarrío-Saavedra, J.; Bossano, R. ILS: An R package for statistical analysis in Interlaboratory Studies. *Chemom. Intell. Lab. Syst.* **2018**, *181*, 11–20. [CrossRef]
- Flores, M.; Naya, S.; Tarrío-Saavedra, J.; Fernández-Casal, R. Functional data analysis approach of Mandel's h and k statistics in Interlaboratory Studies. In *Functional Statistics and Related Fields*; Springer: Berlin/Heidelberg, Germany, 2017; pp. 123–130.
- Liu, R.Y. Control Charts for Multivariate Processes. *J. Am. Stat. Assoc.* **1995**, *90*, 1380–1387. [CrossRef]
- Liu, R.Y.; Tang, J. Control charts for dependent and independent measurements based on bootstrap methods. *J. Am. Stat. Assoc.* **1996**, *91*, 1694–1700. [CrossRef]
- Liu, R. On a notion of data depth based on random simplices. *Ann. Stat.* **1990**, *18*, 405–414. [CrossRef]
- Chatterjee, S.; Qiu, P. Distribution-free cumulative sum control charts using bootstrap-based control limits. *Ann. Appl. Stat.* **2009**, *3*, 349–369. [CrossRef]
- Jones, L.; Woodall, W. The performance of bootstrap control charts. *J. Qual. Technol.* **1998**, *30*, 362. [CrossRef]
- Chakraborti, S. Nonparametric (Distribution-Free) Quality Control Charts. In *Encyclopedia of Statistical Sciences*; John Wiley & Sons, Inc.: Hoboken, NJ, USA, 2011.
- Qiu, P.; Li, Z. On nonparametric statistical process control of univariate processes. *Technometrics* **2011**, *53*, 390–405. [CrossRef]
- Barbeito, I.H.; Zaragoza, S.; Tarrío-Saavedra, J.; Naya, S. Assessing thermal comfort and energy efficiency in buildings by statistical quality control for autocorrelated data. *Appl. Energy* **2017**, *190*, 1–17. [CrossRef]
- Montgomery, D. *Introduction to Statistical Quality Control.*; John Wiley & Sons: New York, NY, USA, 2009.
- Ledolter, J.; Bisgaard, S. Challenges in constructing time series models from process data. *Qual. Reliab. Eng. Int.* **2011**, *27*, 165–178. [CrossRef]
- Box, G.E.; Luceño, A.; Paniagua-Quinones, M.D.C. *Statistical Control by Monitoring and Adjustment*; John Wiley & Sons: Hoboken, NJ, USA, 2011; Volume 700.
- Del Castillo, E. Statistical process adjustment: A brief retrospective, current status, and some opportunities for further work. *Stat. Neerl.* **2006**, *60*, 309–326. [CrossRef]
- Knoth, S.; Schmid, W. Control charts for time series: A review. *Front. Stat. Qual. Control* **2004**, *7*, 210–236.
- Prajapati, D.; Singh, S. Control charts for monitoring the autocorrelated process parameters: a literature review. *Int. J. Prod. Qual. Manag.* **2012**, *10*, 207–249. [CrossRef]
- Tsai, T.R.; Hsieh, Y.W. Simulated Shewhart control chart for monitoring variance components. *Int. J. Reliab. Qual. Saf. Eng.* **2009**, *16*, 1–22. [CrossRef]

- Jones-Farmer, L.; Woodall, W.; Steiner, S.; Champ, C. An overview of phase I analysis for process improvement and monitoring. *J. Qual. Technol.* **2014**, *46*, 265. [CrossRef]
- Grasso, M.; Colosimo, B.M.; Tsung, F. A phase I multi-modelling approach for profile monitoring of signal data. *Int. J. Prod. Res.* **2017**, *55*, 4354–4377. [CrossRef]
- Febrero-Bande, M.; Galeano, P.; González-Manteiga, W. Outlier detection in functional data by depth measures, with application to identify abnormal NOx levels. *Environmetrics* **2007**, *19*, 331–345. [CrossRef]
- Kazemzadeh, R.B.; Noorossana, R.; Ayoubi, M. Change point estimation of multivariate linear profiles under linear drift. *Commun. Stat.-Simul. Comput.* **2015**, *44*, 1570–1599. [CrossRef]
- Woodall, W.H.; Spitzner, D.J.; Montgomery, D.C.; Gupta, S. Using control charts to monitor process and product quality profiles. *J. Qual. Technol.* **2004**, *36*, 309–320. [CrossRef]
- Lanhede, D. Statistical Process Control: Evaluation and Implementation of Methods for Statistical Process Control at GE Healthcare. Master's Thesis, Umeå University, Umeå, Sweden, 2015.
- Flores Sánchez, M. Nuevas Aportaciones del Análisis de Datos Funcionales en el Control Estadístico de Procesos. Ph.D. Thesis, Universidade da Coruña, A Coruña, Spain, 2019.
- Febrero-Bande, M.; Oviedo de la Fuente, M. Statistical computing in functional data analysis: The R package fda.usc. *J. Stat. Softw.* **2012**, *51*, 1–28. [CrossRef]
- Fraiman, R.; Muniz, G. Trimmed means for functional data. *Test* **2001**, *10*, 419–440. [CrossRef]
- López-Pintado, S.; Romo, J. On the concept of depth for functional data. *J. Am. Stat. Assoc.* **2009**, *104*, 718–734. [CrossRef]
- Cuevas, A.; Febrero, M.; Fraiman, R. Robust estimation and classification for functional data via projection-based depth notions. *Comput. Stat.* **2007**, *22*, 481–496. [CrossRef]
- Cuesta-Albertos, J.A.; Nieto-Reyes, A. The random Tukey depth. *Comput. Stat. Data Anal.* **2008**, *52*, 4979–4988. [CrossRef]
- Flores, M.; Fernández-Casal, R.; Naya, S.; Tarrío-Saavedra, J. qcr: Quality Control Review. R Package Version 1.0. 2016. Available online: https://cran.r-project.org/web/packages/qcr/index.html (accessed on 26 December 2019).
- Hyndman, R.J.; Shang, H.L. Rainbow plots, bagplots, and boxplots for functional data. *J. Comput. Graph. Stat.* **2010**, *19*, 29–45. [CrossRef]
- Raña, P. Pointwise Forecast, Confidence and Prediction Intervals in Electricity Demand and Price. Ph.D. Thesis, Universidade da Coruña, A Coruña, Spain, 2016.
- Colosimo, B.M.; Pacella, M. On the use of principal component analysis to identify systematic patterns in roundness profiles. *Qual. Reliab. Eng. Int.* **2007**, *23*, 707–725. [CrossRef]

© 2020 by the authors. Licensee MDPI, Basel, Switzerland. This article is an open access article distributed under the terms and conditions of the Creative Commons Attribution (CC BY) license (http://creativecommons.org/licenses/by/4.0/).

MDPI
St. Alban-Anlage 66
4052 Basel
Switzerland
Tel. +41 61 683 77 34
Fax +41 61 302 89 18
www.mdpi.com

Mathematics Editorial Office
E-mail: mathematics@mdpi.com
www.mdpi.com/journal/mathematics

www.ingramcontent.com/pod-product-compliance
Lightning Source LLC
LaVergne TN
LVHW070609100526
838202LV00012B/599